the Lowing Herd

Alec Bull

Bramerton Jane,
who learnt how to get more grass

First published in Great Britain in 1999
by **Alec Bull**

ISBN 0 9536763 0 7

Design and Production: **Paul Westley**, Norwich, Norfolk
Printed and Bound: **Rapide**, Watton, Norfolk

The curfew tolls
the knell
of parting day,
The lowing herd
winds slowly
oe'r the lea-

Gray's Elegy
written in a country churchyard

CONTENTS

Food For Thought

It was the back end of 1989 when I saw Hector again for the first time in nearly 20 years-and had failed to recognise him. He had put on weight and now wore spectacles, and that mass of black wavy hair was liberally sprinkled with grey. We were leaning on the bulk milk tank watching the milk disappear steadily as the tanker sucked it through the hose.

"I suppose you are one of the last of the old boys."

I looked quizzically at him. "How do you mean?"

"When was it you went to Eglingtons?"

"1960."

"There you are then. In 1960, nearly every farm had a herd of some sort and all the milk was collected by churn lorries. How many herds are there left about here now?"

"Not very many, and most of those that are, are little better than milk factories. Everything is run by computer on a lot of them. The milk goes through a flow meter from each cow, the computer logs it, and the next time she comes in, puts the right amount of concentrates in front of her. The cowman puts the milking clusters on, and that ends his job until next milking so far as that cow is concerned."

"You haven't got anything like that here, I'll bet a bob!"

"No, nor likely to have. I wouldn't want it anyhow. I like to see each cow each milking and make sure she's O.K."

"That's what I said. You're one of the last of the old boys. The young 'uns don't want anything like that today. They want to clock in and clock out and that's it!"

Slurping noises from the bottom of the tank indicated that the milk had all gone. Hector hurried out of the dairy and closed the stop cock. Came back in and unscrewed the hose from the tank, hoisting it back into its channel on the side of the tanker.

"I'll rinse it off for you for old times sake" I said, flicking the pressure hose tap and turning the nozzle into the open tank.

"Thanks a lot. See you tomorrow."

The tanker reversed away from the dairy, swung round and went bumping up the stony track to the A47. I glanced after it. Hector was one of the last of the old boys as well. He had been on the churn lorries in 1960, and here it was 1989. I had not seen him since 1973, when we moved here, as his employers, Wincanton Transport, had not collected the milk from this area, though only a few miles from my last place. Now, under one of the ever more frequent reorganisations of the milk collection industry, Wincanton were going to pick our milk up, at least for the time being.

Dinner time. I set out across the meadow, deep in thought. One of the last of the old boys. I supposed it was true, though it came as something of a shock to have it spelled out so plainly. 1960 was not the beginning of it though.

My thoughts went back to Christmas 1941, when I left school at 14 to work on the family farm and help with the war effort. My first responsibility was to look after the herd of twelve to fifteen cross bred cows. They were milked with a Gascoigne portable milking machine which was about two years old, but they were also hand stripped afterwards as it was felt that the milking machine was not capable of getting all the milk. Because they were accustomed to being hand stripped, they always held some milk back to keep the hand stripper in work, or so it seemed.

It was widely believed at that time that any milk left in the cows udder might give rise to mastitis, which, in the days before intermammary antibiotics was a major calamity. After fifty years of milking cows, I know that more mastitis is caused by leaving the milking clusters on even a few seconds too long, than by leaving a drop of milk in the udder. But then, nobody handstrips today in any case.

I was pitched in at the deep end, because father hated cows, even though he had always had to milk them, even before he left school during the first world war. During the second world war, with the emphasis on producing everything at home, milk was a very important and relatively highly paid for commodity. Virtually every farm had a herd of some sort, sending anything from one ten gallon churn and upwards to the creamery each day. A series of relatively small creameries were dotted about the countryside. Some were in quite small villages, these usually being close to wayside stations on the then extensive railway network. The ten gallon churns collected from the farms were taken to the local creameries where the milk was either bottled for use, probably no further away than the nearest large town, or they would be offloaded into seventeen gallon churns for express carriage to depots in London and other large cities.

I did not mind looking after the cows. It broke up the tedium of field work, having to break off at half past three or four o'clock to come back to the buildings, round up the cows from the nearby meadow and milk them. Milking machines were far from perfect in those early days and, once the cows were in the yard, getting them into the parlour two at a time, feeding them their concentrate, washing their udders and milking them would occupy probably nearly an hour and a half. This may sound a long while today, but after each two cows were milked, the milk had to be poured from the milking buckets into carrying buckets and ferried nearly fifty yards across the farmyard, taken through the kitchen and into the dairy, where it was poured into a receiving tank by way of a strainer. From thence, it ran down a milk cooler which was simply an arrangement of sealed copper pipes going to and fro horizontally, through which water passed in an upward direction. Cold water was essential for this operation, this being pumped from a well under the kitchen sink. How useful, you might say, to have a well under the kitchen sink in those days before the water main ran through the village. The reality was somewhat otherwise. The house was many hundreds of years old and may at some distant time in the past have been connected with a religious foundation. The well under the sink was fed by a primitive pipe which I seem to remember being told, was made of wood and which led to the bottom of the nearby horse pond. The water was therefore far from pure, though in those days it would have had no nitrogen pollution in it. Even so, the pond was partly fed from a ditch which came down the nearby roadside for probably nearly a mile. Apart from cooling the milk, it was used largely for scrubbing the floors in kitchen and dairy which were of white brick. Drinking water had to be carried on a yoke with two buckets from a well rather more than 200 yards from the house. No hosepipe bans in those days! The drinking water well never dried up, though the one in the kitchen must have come close to doing so during summers when the pond became very low. At such times, a horse drawn three-wheeled water butt had to be taken and filled at the village pump, for non-drinking water, as the water in the pump had a high iron content which made it taste very bitter. Drinking water was sometimes fetched using a two wheeled hand water butt from a standpipe which was on the water main, for washing up the dairy utensils, when the level in the drinking water well became low. Both pump and standpipe were a quarter mile away.

As the years passed, I became interested in natural history, starting early in 1944 with birds, though by midsummer of that year, I was also recording plants, butterflies and anything else that attracted my attention. Even until after the war, there were still plenty of old meadows filled with wild flowers which I was fortunate enough to have known before they were destroyed by hedgerow removal and ploughing up. We had several small meadows of our own which at haymaking time, were white with a haze of ox-eye daisies nodding a foot or two above the grass. This too would be coming into flower, and included much crested dogstail and quake grass, the latter known locally as 'ladies hair'. Much of the earlier flowering sweet vernal grass was present in the sward, which is why haymaking in those far off days is always associated in my mind with the glorious hayscent no longer to be enjoyed from todays ryegrass hay.

One of our meadows had a large colony of wood anemones in the middle. This can be taken as an indication of a very ancient meadow, possibly undisturbed since it had been cleared of woodland in the remote past. Our old meadows also had a few orchids, though nothing like the numbers to be found in meadows belonging to near neighbours. A large meadow with a footpath belonging to Elsdens, had a swampy area filled with Green-winged Orchids. From the swamp I would surreptitiously-as I had now left the footpath and was trespassing-follow the right hand branch of the stream and nip over into Le Grice's meadows. There had been four, each of about four acres, but alternate ones had been ploughed during the war. The other two had not. I heard Harry Bugg, Mrs Le Grice's foreman, tell father on one occasion that he did not know what to do with those two meadows. They needed them for hay, but when they cut the hay, they only got a wagon load and a half off each. By that time, father was into ploughing things up and reseeding them, but Bugg did not quite like to do that, so the meadows had a year or two's further reprieve, until the old lady died. End of eight acres of Fragrant Orchids, plus Heath Spotted, Early Purple, Twayblade and even a little colony of Frog Orchids. The last named were rediscovered in 1989 in north east Suffolk, having been believed to be extinct in the county for some thirty years.

Towards the end of 1945, I started corresponding with a girl a few years my senior, who was working as a tractor driver on a farm near Clacton on Sea, Essex. She too was keen on natural history and we had both discovered dog roses flowering in October, as did a number of other people in that year, resulting in a number of letters to the East Anglian Daily Times. During the following winter, we met two or three times for bird watching trips along the Stour estuary, or to Abberton Resevoir south of Colchester, for which Rita had a permit.

Time does not stand still and by early 1948 we wanted to get married. She was 24 and I was 20. My parents opposed the match at first but eventually capitulated, though it meant that we should have to move away as there was no cottage attached to the farm. It so happened that the dairy herd at home was contracting due to a number of cows contracting contagious abortion, later to become more widely known under its 'official' name of Brucellosus. We never knew where it came from, though there were two possibilities. One was the stream running through the cows meadows and from which they drunk, though the more likely source were the cows brought for service to our bull from surrounding small herds up to two or more miles distant. These came sometimes on a halter with a single handler, more often loose with one man in front and another behind. At that time, all roadside hedges were stock proof and with closed gates. By 1948, we had switched from the bull to the Milk Marketing Board's artificial insemination service, then in its infancy, which operated from Beccles. Whether this was as a result of advice from the vet following the abortion outbreak, or stemmed from an accident in which father was bundled through a hole in the rails of the bulls pen which he had gone in to repair, I cannot remember. I do remember that he was knocked unconscious and woke up in the next pen with the bull breathing fire through the hole in the rails, and would undoubtedly have been killed had he not been thrown clear as no one was within earshot at the time.

With the small herd decimated, it was only a short step to reduce it to a single cow for the house.

Soon after we were married, a job was advertised in the "East Anglian" for a "General farm worker to milk house cows and look after other stock, drive tractor, one able to stack and thatch. Wife to help in farmhouse when required." This was at Icklingham in the Suffolk part of Breckland and twenty miles from Hitcham. We both went for the interview and were offered as accomodation either a flat in the farmhouse, or a very isolated flint and brick bungalow with a potentially large garden. The farmer was cultivating part of it with a tractor and plough that year and had potatoes on it, though we should be able to crop it next. In the garden was the remains of a brick kiln, and at the far end, a large, bush grown pit from which the former brickmaker had dug his clay. Whilst looking round, we spotted a cock Redstart flitting alongside one of the rambling hawthorn hedges which surrounded the property on three sides. That did it. We opted for the bungalow and moved in on June 9th 1948.

Icklingham

With present day examples of brief, tempestuous and unsuitable marriages entered into without due thought, not to mention casual liasons with no intention of making them last, ours would seem to have been doomed from the word go. We had corresponded for just over two years, but had only met about a dozen times, and for never more than three or four hours, and we had not always seen eye to eye when we had met. This was largely due to the fact that I knew virtually nothing about the foibles of the opposite sex, having spent most of my time living and working on the farm. I had belonged to the Young Farmer's Club in Stowmarket, whose meetings I attended most weeks during the winter, as well as those summer events that were within cycling distance on a summer's evening. I also indulged in birdwatching trips when time allowed, though usually, I had to get back to do the afternoon milking.

Now, here we were, setting up home with total assets of less than £100, most of which was spent, along with some furniture dockets that Rita had obtained. Along with most other things, furniture was still rationed in 1948. For my weekly wage, I was to receive £3/19 shillings per week, to rise to £4/4/0 when I was 21 in two and a half months time.

We had been over and spent a day cleaning up the cottage which had stone floors, all of which needed a thorough scrubbing. The water for scrubbing had to be laboriously wound up from the depths of a well in a five gallon bucket. Such a large winding bucket was necessary as it was sixty feet down to the water, and a lesser bucket would have lost too much of its contents as it swayed back and forth on its long upward journey. We learnt later that at Lodge Farm Cottages, which were two miles further up in the Forestry land than we were, the well had a more or less endless well rope with a bucket on either end. The well was double the depth of ours so, as the empty bucket was wound down, a full one would be coming up on the other end of the rope. For some reason, I did once lose one of our own three gallon buckets down the well. I borrowed a three pronged grappling hook known locally by the unlikely name of 'the creepers', and went fishing with it on the end of the well rope. After an hour or so, I had a haul of six buckets, of which four including the one I had lost, were sound and in good condition.

Having raised the water, it had to be taken to the copper in the wash house thirty yards away, and heated with wood collected from the nearest pine belt, of which there was one round almost every field.

On this first occasion, I hauled the water and put various other jobs in hand while Rita did the scrubbing, and we made our way back to Hitcham. To shorten our journey, we had taken our bikes on the train from Lavenham to Bury and cycled from there, though this only cut the cycling distance by eight miles. When it was time to go, my bike had got a puncture, so I had to hitch hike back. Much less traffic in those days, but strangely much easier to hitch a lift. The next morning, I borrowed a bike to get to Lavenham and left it at the station, hitching from Bury to Pear Tree Cottage, and got out from my third and final lift at the bottom of the track as Rita was disappearing up it. This last lift had only been a short one but proved to be very useful, as it had been in the bread van driven by Cecil Bell of Chimney Mill, West Stow who readily agreed to deliver bread to us and, together with Mrs Fish, the milk lady from the same village, constituted the only callers we had.

Pear Tree Cottage, Icklingham

Pear Tree Cottage stands beside the Icknield Way, the old arterial route into East Anglia from the far south west of Britain which has been in existence since Neolithic times at least. Crossing the River Lark at Lackford, it runs east of north for nearly five miles in a straight line as a sandy track until it reaches the Bury to Brandon road at Shelterhouse Corner. A quarter of a mile from Lackford Bridge, the Way crosses the West Stow to Icklingham road at the corner of Letch (pronounced Leech) Moor. Here were a pair of cottages, since put into one, who were our nearest neighbours. Pear Tree is another quarter of a mile towards

Icknield Way crossing West Stow road, Letch Moor on the right

Elveden. Our most intimate neighbours were the rabbits that lived in the brick kiln and clay pit beyond, not to mention those in the former sandpit on Forestry Commission land just across the track. To be able to live, we had to grow vegetables and to do that, we had to wire net an area that we could keep an eye on close to the house. Potatoes we could grow on the plots, which, for the first year, were already planted. Rabbits did not eat potato tops, though they would gladly dig the tubers out once the tops had wilted a bit.

What about other necessities like groceries and meat? There was a grocers shop in Icklingham, two miles away, which sold most things that were available at that time, so it was to Miss Hurrell to whom we took our ration books. She, dear lady, was very kind to a newly wed couple just set up in our first home. She recommended that we should have Mr Gentle from Brandon as our butcher, and if we gave her our order each week, she would take our meat in for us and keep it until we could collect it. Neither of us ever met him, but even now I remember his sausages as being among the best that we ever had.

All our food came from Miss Hurrell, but to heat the oilstove that Rita had for cooking, we had to go to the other shop in the village to buy paraffin from Mr Brown. A haversack full of groceries, with a gallon can of paraffin on the handlebars was not easy to manage, or very suitable companions, though I cannot remember any catastrophies.

In her younger days, Miss Hurrell, who was broad and rosy-cheeked, with white hair and a spotless white apron, had been cook to Mr Hore-Belisha, the Minister of Transport who gave his name to 'Belisha Beacons', still in use at either end of Zebra crossings. She told us on one occasion that Mr Hore-Belisha had often said that he would have to get married if she left him, and that was what had happened. Her brother was postman, and left our letters to be collected from the farm. Our newspapers were also left there which was annoying as we had to collect them whether I was at work or not, and we still had to pay the full delivery charge even though the papers were delivered with the Majors'.

Having got ahead of myself, I will return to Pear Tree Cottage on the day of our arival. The Hitcham carrier had brought us some items of furniture donated to us by family, but the bed, wardrobe, dressing table, drop leaf table and four dining chairs on which the dockets had been used, were due to arrive from Duttons of Bury St.Edmunds. They did not. We waited all day, and I finally cycled down to Weatherhill Farm at ten to five to ask Mrs Barnicot, the Major's cousin and housekeeper, if she would telephone them for us. She did, and gave them a good roasting when they said that they were coming next morning. They were very apologetic and the lorry arrived at seven o'clock. By the time everything was in place we were more than ready for a comfortable bed to sleep in that night.

Major Lucas, my new employer, was an unusual farmer even as military gentlemen go. For one thing, he had very little military presence about him. He had obviously spent a large part of his life abroad, as even after many years of 'retirement', his complexion was still that of mahogany. He was tall and spare, with a shock of white hair, a longish, thin face and far seeing blue eyes. At 70 he only wore glasses for reading. He had been in the Army Medical Corps, spending much time in India during the Edwardian period, being twice Army tennis champion out there, and twice, Master of the Poona foxhounds. Therein lay his interests. His farm had been placed under a tenant farmer for the duration of the war by the War Agricultural Executive Committee, and when we went there, he was regaining possession the following Michaelmas Day. He already had most of the farm buildings, one or two immediately adjacent paddocks and curiously, two or three one acre strips of Jerusalem Artichokes which were dotted about the farm. Ostensibly, these were for sale when the price of artichokes justified lifting them. When it was not, they were fenced so that his herd of Large White breeding sows could be turned into them to root as many up and eat as they could find. By this method, there were always enough left in the ground to provide for next year's crop.

The real thinking behind the artichokes was that they made excellent cover for pheasants which were, and always had been, the Major's chief interest. To help him look after the pheasants, he had five golden labradors and two French poodles, one of which was his inseparable companion. Cocoa was cocoa coloured, quite large and with tight curls all over. None of your ridiculously trimmed and pampered pets here. She had an amazing turn of speed when retrieving game and had, as the Major never tired of telling, the gentlest mouth for a partidge as any dog he had ever had.

He also had Herbert. Herbert was a German prisoner of war who lived in a one roomed flat in the farmhouse. Strangely perhaps, in such a situation, he was still a fanatical Nazi who had seen service on the Russian front, and suffered kidney trouble as a result of privations during the Russian winter. When they had fallen back ahead of the Russians, his unit had been captured by the Americans and he had initially been sent to the United States.

I do not remember how he came to be back in Britain. He had been sent to the Major by the War Ag. from the large prisoner of war camp that was at Mildenhall at that time. He lived in and was acting in the capacity of foreman on the Major's part of the farm. This was a bit uncomfortable at times, but with labour and machinery both scarce, only he could handle the parties of Ukrainian PoWs who came on a regular basis when Michaelmas had come and gone. A few of these had a little English, which Herbert spoke quite well, and they all understood German. All the prisoners wore the regulation uniforms with dark, round patches stitched on the knee and back so that they would be instantly recognised if they made a break for it. By 1948, most had no incentive to do so as repatriation was already under way. In any case, few wanted to go back to Stalin's Russia, and many settled permanently in this country. Herbert himself went back to Germany late the following winter, his wife having managed to reach the west from their pre-war home in Breslau, then in the Communist East. He had been a master builder before the war, and had built the brick pig styes that housed the Major's pigs, a brick pit in which small surplus potatoes could be ensiled after they had been boiled, and the twin coppers which stood to one side of the yard, in which the boiling was done. He also looked after the dogs.

Bridge Farm, Lackford was at that time farmed by a very elderly farmer, J.D.Sayer who, in his middle 80s could often be seen driving round the farm and its vicinity in a pony and trap. Always known as

'J.D.' he kept a large flock of Suffolk sheep, some of whom looked nearly as old as their owner. "Bull! Take the tractor and trailer and go to J.D.Sayer's and pick up a couple of dead sheep will you?" became a familiar order over the next few months. Soon, I would arrive back and dump the carcasses under the pine belt by the dog runs. Herbert would string it up in a tree, out with his skinning knife, and would have its 'jacket' off in no time. The skins were usually roughly cured and would be sold if good enough. If not, they came in very handy for softening the impact of the hard metal bucket seats of the tractors, or for wrapping round the tractor driver's knees if he were at plough on a cold winter's day. No centrally heated cabs in those days.

On one occasion, I was surprised to find a horned, black faced sheep to be picked up. No one seemed to have heard of Suffolk sheep having horns. It was not until 1993, when Peter Wade-Martins published his book 'Black Faces' (Published by Norfolk Museums Service) that I realised that the apparent anachronism was a Norfolk Horn. J.D. had been keeping the old Norfolk breed alive single handed for more than 30 years, though the numbers had fallen to a mere handful by 1948, and it was from his stock that the breed was rescued. By careful back crossing with Suffolks who are directly descended from the Norfolk Horn, a new Norfolk Horn Sheep Society has managed to get a once more thriving, though still very much a rare breed, back from the brink using the last of J.D.s lines.

Once skinned, the carcases were cut up as needed. If there were a glut, they might hang in the pine trees for a fortnight, and there were gluts at certain seasons as the major cause of death was from liver fluke, a parasite transmitted to the sheep inadvertantly by way of a snail of damp places such as those alongside the River Lark. This did not matter during the winter, but during the summer, the farmyard had its own aroma. When pieces were wanted, Herbert cut them off and hung them in the meal shed ready for chopping up in dog sized chunks. During the summer, an injudicious jostling of a hanging joint whilst getting cattle or pig food, resulted in a neck full of juicy maggots. The entrails were all boiled up in the coppers, the pluck being fed to the dogs and the intestines and resulting soup going to fortify the pig food.

Vladimir Lpunka was the one regular among the Ukranians. He was large, strong as a horse, and absolutely obedient to Herbert but, in local parlance, he was a bit 'thick'. It was often his job to light the coppers when he had been dropped off the truck first thing in the morning. One morning, he was busy doing this while I was doing something round in the main barn. Suddenly, there came a tremendous muffled explosion somewhere round the yard. I ran round to see what had happened and arrived in time to see Vladimir crawling out of the back of the cartshed a full twenty yards from the coppers. He had lost his eyebrows and the front of his hair was badly singed. It appeared that the coppers were not lighting very well that morning. In fact, he thought they had gone out. They had not. The flames had died down, but there had still been a glow. Vladimir had got himself a tin full of tractor vaporising oil, on which most tractors used to run before diesel became the driving force for most farm machinery. He had hurled the whole tinfull into the copperhole and the TVO vapour had exploded hurling him the twenty yards into the back of the cartshed. I never could fathom out why he was only bruised. Herbert had to repair the coppers as the housing was cracked- right across.

I wondered at first why the Ukrainians all looked up to Herbert as they did. They did as he told them without question, and kept him supplied with goods 'borrowed' from the camp which he somehow sent over to Germany to friends and relatives, either for their own use or to sell on the black market. His room was often stacked high with government issue ersatz coffee as well as less noticeable items such as saccharin. On one occasion the old postman, an ex-sergeant major from the first war, with long waxed moustachios, and Miss Hurrell's brother, showed me Herbert's current mailing, a pile of letters to addresses in Germany all with a thickening in the middle which rattled, and would probably have contained fifty or more saccharins. I eventually discovered the reason for the Ukrainian's apparent devotion to Herbert during a conversation with Vladimir on one occasion, whose English was passable. He told me that he and his countrymen had fought with the Germans because "Communismus no good!" He told me that his parents were in a concentration camp in Siberia with over two million people in it. The number could certainly have been an exaggeration, but we now know that many people from Russia and its provinces were sent to Siberia under Stalin under the slightest provocation. The day after Herbert went home on leave with the option of staying there or returning to the Major, Vladimir fell ill, and stayed that way until the Major heard that Herbert was not coming back. Then he too said he was not coming back. Looking back, I think this may have been the time when the PoW camps were closed, as we never had any more on the farm after that date which was late January 1949. This left me with the only one other person on the farm to run the whole show for the Major on over 300 acres until two new workers started after the week end of April 24th when they moved in. This included all the spring cultivations and also some threshing. The only other person was Arthur Blyth from Letch Moor cottages, who did the kitchen garden and was general dogsbody round the place. He was 74 at the time, and though bowed, could still do a good days work, though he started at 9.00 am instead of 7.30. He was short, with wrinkled features, pale blue eyes and a straggly moustache which was usually stained with tobacco juice, as he chewed incessantly on his toothless gums. He was known as 'Kronje' to his contemporaries, his war having been against the Boers.

During those three months, apart from milking the house cows twice a day, which I did, Arthur took on the feeding of the pigs and calves, and the general yardman's work which included skinning sheep and boiling coppers.

In addition to a certain amount of bought in meal from a local millers', and boiled potatoes, large drums of Tottenham Pudding so called, arrived at regular intervals to supplement the pig food, as cereal based animal feeds were scarce and still strictly rationed. Tottenham Pud had been invented during the war, being made from collected kitchen waste in our towns and cities. It was boiled and drained, compressed into large, barrel shaped metal containers and sent to farms specifically as pig food. It was no doubt jolly good stuff, but we had to watch carefully what was mixed with it. Old razor blades were quite frequent and, if fed to an animal, could have caused considerable damage. On one occasion I found a very new looking and perfect bottom denture. I cleaned it up on a piece of sacking and took it along to Arthur.

"Arthur, have you lost your bottom denture?" That in itself was a joke, as I never saw him with any teeth at all. I held them out.

"No bor, they en't mine. Mine are at home in a jam jar on the mantelpiece." He held out his hand and took them. Turned them over two or three times.

"No. Don't know whose they are." Before I had a chance to say any more, he suddenly opened his mouth and popped them in, chewed them round and round reflectively for a minute or two, then spat them out again.

"No! They definitely en't mine. Still, I'll take them an' see if I can find who they belong to if you like."

Whereupon he popped them in his pocket. I did not have the heart to tell him that he never would.

Toward the end of winter, it came to be time to thresh the stacks of wheat I had built last harvest. Because we were short handed, I had to go onto the cornstack with two women from the village to pitch the sheaves onto the drum, while Arthur was set to stack the straw behind the baler. Until that time, a straw pitcher had been traditional, with the straw being stacked loose, but this threshing set had one of the early balers that tied the straw in large, heavy bales using wire instead of string. Arthur had never stacked bales before. Nor for that matter, had I, but he seemed to be getting on alright, and by leaving off time, the walls of the bale stack were nine or ten feet from the ground. As he did not start until 9.00 o'clock, I had to stand in on the bale stack until he got there. All went well until I reached one of the corners, when I suddenly felt an outward movement. I threw myself into the middle of the stack just in time, as the whole corner toppled over right down to the ground. Arthur had been stacking the bales two one way and then two the other all the way up, instead of overlapping them alternately in each direction. This meant that the whole stack was a series of slender columns two bales each way right down to the ground with no visible means of support.

The wheat which had been put into the stacks came from the newly broken up riverside meadows, and the sheaves had contained as much grass in places as there had been wheat. It was traditional that wheat sheaves could be carted the day after they were cut, due the nature of the grain and providing the weather was fine in the meantime. So the Major decided to cart behind the binder. I thought it very risky and Herbert, who was still foreman at that time, demurred, but the Major had had his way. When we took the stacks down, they were all a rich brown in the middle, having heated almost to the point of spontaneous combustion. This frequently happens with hay that is carted before it has properly 'made'. The wheat sample which resulted was of very poor quality!

The Major never had any hay catch fire to my knowledge, as one might have expected from the foregoing. He was too canny for that, or should I say that he was aware of certain shortcomings which he had in respect of agricultural practice. Hay would be cut, left for perhaps two nights before turning then carted next day. Even during a hot dry summer on the Brecks it would not have been fit for carting in that short time, so, to prevent combustion, the Major had the stacker sprinkle a thick layer of salt between each layer of grass to keep things cool.

From the time of stacking to the time when all the stacks had been carted away for litter, the Major would inspect the yard and stackyard daily. If so much as a couple of straws were to be seen it would be

"Bull!" pointing to the offending debris "An untidy stackyard is the hallmark of a bad farmer!" I wonder what he would say today if he could see the long lines of partially sodden and rotting round bales stretching for hundreds of yards along field margins?

He also had a keen approach to business on a small scale. He would go to a farm sale nearby and make a number of purchases, usually worn out implements that were the despair of Herbert. On one occasion, he had been out, and on his return, he came across to me in a mood of suppressed excitement.

"Hitch the tractor onto the trailer and get Herbert and Vladimir. I have bought some things from a sale at Flempton. I have bought another Ferguson tractor with plough, ridger and steerage hoe. Also some pig arks. The tractor and its implements cost me £125!" He sounded awed at his own audacity. When I told Herbert, he did not believe it.

"A hundred and twenty five pounds? Ze Machor never pays more zan fifteen shillings for anysing."

Still, he had to accept that another tractor and its equipment appeared to be coming our way.

Off we went to Flempton, three miles away, being passed by the Major in his 1929 Morris ten on the way, his old brown trilby pulled well down over his eyes as was his wont. When we got to the saleground, he came across and indicated the pig arks which we had to dismantle and load onto the trailer. They had cost him fifteen shillings. Presently, we noticed him in close conversation with another farmer near the tractor. After a long time, he came across.

"You can load the steerage hoe onto the trailer as well. That is what I wanted. I have sold the tractor, plough and ridger to Mr X. for £125."

He went off chuckling. Herbert watched him walk away. "I told you he never pays more zan fifteen shillings for anysing!"

When we got back to the farm, the steerage hoe was unloaded into the barn where it was to spend the winter. Some weeks later, the Major was directing some alterations in storage which involved moving the hoe in an arc so that it was standing back against the wall. Vladimir grabbed one end and started to do just that, the hoe blade at the far end making a nasty screech on the floor. The Major winced visibly.

"Oh! Do be careful of that! It cost me £50."

There were two things at which he did excel. One was as a sportsman, that is, with a gun, though he had excelled at other sports earlier in his life, and the other was as an amateur archaeologist.

He looked after his pheasants well, though he did not rear any as that was then not acceptable, with feed so scarce. When he had a shooting 'party' just himself and two friends, each with their own dogs walking the whole of the 300 acres, they bagged just over 80 pheasants in the day, and most of these would have been cocks as the hen birds were often spared to provide for the next generation. Pheasants are polygamous and may have up to half a dozen 'wives'. Too many cocks on a shoot would lead to too much competition for the hens and a probable lower success rate of reared chicks.

To pinpoint the Majors prowess with his twelve bore there follows what might be described as a 'winter's tale'. Dressed, as ever, in his old brown trilby, and with a shabby overcoat draped round his spare figure, he set out from the farmyard carrying a box containing a ferret and with his gun under his arm. Disappearing into the field pit behind the barn, he settled himself on his inseparable friend, his shooting stick, having popped the ferret into one of the honeycomb of rabbit holes in the banks of the pit. From time to time, a white face would appear at the entrance of one of the holes to have a breather and as quickly disappear again. Sometimes a rabbit would dash out and scramble for the safety of another hole, perhaps no more than three feet away. The twelve bore would come up like a flash, and with scarce time to aim, bunny would be kicking his last. The Major returned to the farmyard and came across to me.

"Bull! Take the tractor and trailer will you and collect the rabbits? There is also a rat, which the ferrets can have."

I loaded seventeen rabbits and the rat onto the trailer and strolled across to where the Major had been perched on his shooting stick. There were eighteen cartridge cases.

Archaeological friends were frequent visitors to Weatherhill Farm, where the showcases and drawers in his study were a constant source of interest. All the specimens on show had been picked up on the farm and ranged from a large oystershell bedded in flint to some of the finest flint arrowheads that I have ever seen. There were also many Roman artefacts, some of which came from one of the fields known as 'Horselands', the known site of a Roman villa which had never been properly excavated, while others came from the 'Back of the Barn Field' where the Major himself had done some excavation of a Roman pottery kiln associated with one of the two pits from which the potter's clay had been taken.

I ploughed and cultivated on Horselands a number of times while we were there, and though I spotted pieces of pottery and hypocaust brick, one of the commonest of Roman remains, I never found anything of greater interest. Not that I expected to, knowing how the Major walked carefully over that part of Horselands each time it was cultivated. When he died only two or three years later, the farm was left to his cousin, General Browning, whose son John has had a long battle, trying to obtain the return to Britain of the Icklingham treasure from a New York art gallery, where it arrived by devious means, having been discovered illegally and stolen by metal detectors working under cover of darkness.

Berner's heath, Icklingham, April 1974

Pear Tree Cottage

What of our life in the wilderness? How were we coping with shortages, difficulties with not only obtaining, but also hauling supplies? In time, there was also the baby that was coming. What about the loneliness? Apart from Cecil Bell and Mrs Fish and the occasional forestry worker cycling past, no one ever came near Pear Tree Cottage. The view from the kitchen window embraced the rabbit infested garden as far as the bush grown clay pit at the far end. At the front, there was the grass edged sandy track that was the Icknield Way, and beyond it, the bush grown small gravel pit behind the forestry fence. Beside that was a large derelict area, scrubby with brambles and gorse and one or two small oaks where Jays had planted acorns in the past and never returned to dig them up again. Two or three hundred yards away the Cadogan Estate, the owners until the death of Lord Cadogan in 1935, had ill advisedly planted a sycamore plantation at some time. This had grown to an age of probably thirty to forty years, when all the trees seemed to have died more or less simulaneously, probably from drought.

Following the death of the Earl, the estate had been bought by the Forestry Commission in 1935, the year of the Silver Jublee of King George V and Queen Mary, and was christened the 'King's Forest' as a tribute. The Cadogan Hall and Park were bought by the Commissioners of the Methodist Council and became the now famous Culford School.

Above and below the bungalow were arable fields belonging to the Major. Diagonally across the Icknield Way on the lower side of the old gravel pit was a large field at that time owned by Mr Wagstaff from West Stow. Invisible from the house, beyond our bush grown clay pit and in the next field, was the former parish sand pit. In 1941, after Pearl Harbour, the Americans came over in a big way, and the Eighth Air Force were deployed all over East Anglia. To accomodate them, airfields had to be built in double quick time. Airfields meant concrete runways, and concrete runways meant large quantities of ballast as nearby as possible. One of the projected airfields was at Lakenheath, where the nearest possible source of ballast was the old parish sandpit at Icklingham. Having nothing else to do, with the remainder of his farm in the hands of a tenant, the Major spent April, May and June of that year logging 10,000 four cubic yard lorry loads of ballast as they were hauled away to Lakenheath by H.Allen Newport, a local contractor. The tally books were still in a chest in the loft over the

Major's garage in 1949.

The resulting slash in the side of what had been a grassy knoll was remarkable. Probably 60 yards across and 150 from end to end, the walls fell sheer for more than sixty feet, the depth of the gravel bearing deposit above a thin band of boulder clay which formed the water table, and was level with the bottom of our well. The gravel was not of particularly good quality, as there was a fair amount of chalk outcrop amongst the sand and stone, presumably why there has been no further exploitation since the war. The pit was on two levels, and by scrambling down a fairly steep slope beyond the bottom of the garden, one could peer round a heap of spoil into the bottom of the main pit which, for a large part of the year, held a smallish pool of water in the clay basin. A good place to watch at any time. A secluded pool such as that was a congregating point for birds needing to drink and bathe in a land where water was always in short supply. Thirsty Crossbills and Redpolls, the former from feeding on the seeds of pine in the numerous belts, the latter also seed eaters from larch and birch plantations and clumps which dotted King's Forest, even so early in its existence. During the migration season, they would sometimes be joined by a passing Green Sandpiper. These birds, dark above and white below and on their rumps, gave the appearance of huge House Martins when disturbed, a flash of black and white disappearing out of the pit entrance uttering their characteristic "Tlooi-tlee-tlee".

In the late forties, the cliffs were still sheer and relatively new. Ideal nesting places for Sand Martins. We counted over 300 nest holes in 1949, all in use. Rabbits had burrows, even along some of the narrowest ledges which they could only reach by a very steep scramble, and some of these were regularly appropriated by Jackdaws, of which , three or four pairs nested every year. The area was rather short of hollow trees or suitable farm buildings, and not only the Jackdaws nested underground, but also one or two pairs of Stock Doves, whilst the oddest neighbours of all for the rabbits were a pair of Little Owls that nested successfully in an isolated burrow at the top end of the pit.

We had built a rough bird table and put it halfway between the apple and pear trees in the garden, in full view of the kitchen window and about twenty yards away. A few birds came to it during the winter, though the winter feeding of birds was very much in its infancy, and birds were not in the habit of being fed. In any case, there was not too much food to spare for whatever feathered friends we had round the garden. It was during the hot summer of 1949 that our bird table really came into its own, for on it we had a flat pilchard tin which was kept full of water. During that summer, we listed 48 species of birds on the bird table, most of which had dropped in to drink, or bathe, or both, sometimes two or three times during the course of the day. The most frequent bather was our gorgeous cock Redstart. A flash of fire as he landed on the edge of the bird table with a 'wee-tic-tic", a quick hop and iridescent drops of water would be flying in continuous spray for perhaps two minutes before he even came up for air, the white slash gleaming between his blue-grey head and black cheeks. His bronze breast would fluff out for a quick shake of glistening droplets, a fiery flick of his tail and he would disappear into the unpruned branches above to have a private preen, a brief snatch of scratchy, queruolus song, and he would be off about his business for a few more hours. Strangely, we rarely saw his sober coloured wife at the tin.

The 'tinies' came to drink and bathe as well, preferring a lower water level than the Redstart. The Warblers and Tits, and occasionally the smallest of all, the Goldcrest. The Tree Creeper, unused to being on the horizontal, adopted an unusual method to have his bath. He would perch on the tin facing outwards, dip his tail in the water and flick it over his back, accompanied by much shuffling of feathers.

Fortunately the Red-legged Partridges who landed on the table from time to time, only stopped for a sip or two. On the other hand, the Sparrowhawk occasionally seen surveying the table from a low branch, appeared to be searching for bathers rather than bath water!

The washing line was stretched across the front of the house between two rather seedy spruce trees, the best site for catching wind and sun as it was not within the confines of the high hawthorn hedges. One morning, we woke early to hear a great "Chuk-chuk-chukka" going on somewhere nearby. In those days, it was quite safe to leave the washing out all night, or even to go out for the day without locking the door. On this occasion, a full line of washing had been left out, stretching from one tree to the other. We crept out of bed and peeped round the curtains and were amazed to discover eleven Red-legged Partridges spaced out among the washing, chortling to the early morning sun.

This was not as disastrous as an autumn occurrence in which the washing was liberally plastered by a big flock of Starlings who had been indulging themselves nearby on a large crop of elderberries!

A different 'washing' story arose from an occurrence on April 5th 1949. It was a clear, cold and sunny day with a half gale blowing from the south west. An ideal day to wash the heavy cotton bedspread and get it dry, thought Rita. I was harrowing on the field below the bungalow, and kept noticing that the fields on the other side of the river, about half a mile away, were blotted out at frequent intervals by squally dust storms. Breckland in spring has always been notorious for sand storms, one of the most notable of which was in 1665, when a gale whisked up large quantities of sand from Lakenheath Warren, then a bare open space, and deposited it five miles away at Santon Downham, completely blocking the river Little Ouse. This was not quite as bad as that. Most of the wash dried very quickly and was taken in again, but the much thicker bedspread was left a while longer. When Rita went back to get it in, she found that she could hardly lift it down off the line. It had become so deeply ingrained with sand particles that it took several changes of water to float all the sand out. Next day, I saw exactly what damage blowing sand could do. The Major had a field of newly emerged spring oats on Chalkpit Field which was about six inches high. Halfway up the west side was a gap in the hawthorn hedge about fifteen yards wide. The sand grains funnelled through this gap cutting the tender oat shoots off at ground level, and by the time the far side of the field was reached, the fifteen yard swathe had become fifty, with nothing to harvest from the cut off patch at all.

Ramparts Field. Spring 1985

Today, as one leaves the A1101, Bury St.Edmunds to Mildenhall road, to drive to West Stow, after about 100 yards on the left is the car park for Suffolk County Council's Ramparts Field picnic site. A pleasant spot, with pine, spruce, larch, oak and birch on a knoll and with lots of gorse and heather. The flat area on either side of the car park is shortly grassed, and ideal for ball games and exercising the dog. The western end, free of heather, is chalky, and supports a number of local and scarce plants, with the rare Breckland Thyme, Maiden Pink and Glabrous Cats-ear mingled among the commoner species. Across the line of the Icknield Way, no longer signposted as that other than as a footpath, though it was at the time we lived there, one enters a dark aisle, with Corsican Pine reaching to the sky on either side of the road for the next quarter of a mile. A terrace of cottages on the left is opposite the entrance to West Stow Country Park. Again, there is ample parking, an adventure playground and a visitor centre where dioramas show the history of the area, and especially of the excavated and reconstructed Anglo Saxon village which lies at one end of the park, entry to which is on payment of a small fee. Much of the park is grassed and fenced and supports a liqourice all sorts flock of supposedly rare breed sheep. Certainly there are Soay and Jacobs and others, and hybrids. These feed over a long grassy ridge which, with paths and a series of gates, allow visitors to stroll among them. At the east end, and at a lower level, there is a curious looking Victorian building with a tall chimney at one end, and nearby, facing across the River Lark, is a birdwatching hide which looks out onto the Suffolk Wildlife Trust's Lackford Pits Wildfowl Refuge.

Former sewage farm pumping station, West Stow, 1985

How has this countryside changed since 1949? A few years ago, I took a party of naturalists for a walk round the area, culminating with a visit to the Anglo Saxon village. Looking all around him as we strolled among the grazing sheep, I heard one man say to his wife, "Isn't it marvellous to be able to see Breckland as it used to be?" I chortled inwardly. That area has changed more than any in the past half century.

In 1948 and 1949 , we used to sneak in and creep round with our field glasses. Rita actually had a pair of 8x30 prismatics, though I had to be contented with some battered, straight vision field glasses with a magnification of about x4. Our quarry were the many waders that frequented the area, especially during the migration season. Eventually, we were able to obtain a permit to visit the sewage farm legally from the Bury St. Edmunds sewage dept. The Victorian building was formerly the pumping station which brought the sewage through a yard high cast iron pipe from the ancient tunnels that honeycombed the streets of that town. The pipe emerged from the ground about fifty yards from the building, where it crossed a squidgy ditch and made a handy bridge from the open heath onto the lower level of the sewage farm which was out of sight from the cottages, where the sewage farm foreman and his assistants lived. After passing through the pump house with its various filters, liquid from the crude sewage was drained out onto a series of lowbanked 'fields' , each between one and two acres in extent. As each field became filled, the effluent would be channelled to the next, and the water allowed to drain naturally away into the sandy soil, from whence it found its way into an evil smelling stream always known to us as the 'sewage ditch'. This ran parallel with the River Lark for about half a mile, before actually emptying into the River Lark just upstream from Lackford bridge. By this time the effluent was adjudged to be pure enough not to kill the fish, and certainly I do not remember seeing any dead fish in the river during the time we were there. Settlings from the pools were removed periodically and were much sought after by surrounding farmers as a cheap form of humus for the impoverished soils of Breckland. Needless to say, the Major was among those who used sewage sludge.

When a pool became thoroughly dried out, it was ploughed. This was still being done with a pair of horses when we first knew the farm, but a switch was made soon after to a grey Ferguson tractor with a hydraulic lift plough. While the pools were draining and drying out, a continual succession of wading birds would drop in for easy pickings, and, over the years that we were within reach, having got our official permits, we saw many species. Green Sandpipers were the most frequent, both as a passage migrant, and also, one or two usually stayed for the winter. Usually solitary, on one pool particularly favoured by the species, we once counted 52 together. Common Sandpipers, Ruff and Greenshank were all regulars on passage, while Snipe, present all the year round as they bred on nearby marshy meadows, also spent the winter in goodly numbers. Redshank too, were summer visitors breeding on the same meadows with the Snipe and Yellow Wagtails, and feeding with them on the pools, their numbers augmented in August and September by others on passage. Less common waders sometimes included the elegant Wood Sandpiper, longer on the leg and more graceful than its two cousins. Though diving duck would not have been able to feed on the pools, dabbling duck were numerous. It was here, soon after we went to Pear Tree Cottage, that we first made the acquaintance of Gadwall. The duck of this species is scarcely distinguishable from that of Mallard, other than by the white speculum, the coloured panel in a duck's wing which can be seen when the bird is at rest, and appears as a block near the trailing edge of the wing when in flight. The drakes are very different, being a sober, slaty colour with dark grey-brown head and black under the tail, noticeable when the bird 'up-ends' while feeding, when compared with the iridescent green head and pale whitish undertail of the Mallard drakes, the 'Wild Duck' of artists. Gadwall have always been winter visitors to Britain, but were introduced as breeding birds to estates in Norfolk and Suffolk during the 19th century when pinioned birds were turned down and soon became part of our breeding duck fauna. Today, they are common as breeding birds in East Anglia and parts of Scotland and Ireland. While the East Anglian birds were introduced, the Scottish and Irish birds are considered to possibly be natural colonisers as birds from Iceland have always been known to winter in Ireland, presumably visiting Scotland on their long journey south.

The other two dabbling duck commonly breeding locally in the forties were Shoveler and Teal. Whilst looking after the cattle on the Major's marshy meadows, I found nests of each of the four ducks with the exception of Shoveler. Though the sewage farm has long since gone, gravel workings just across the river which have been returned to nature bit by bit as they were worked out, and turned over to the Suffolk Trust for their wildfowl refuge, have enabled diving duck to colonise the Lark valley, with Tufted Duck and Pochard both probably breeding, wintering in good numbers, and sometimes accompanied by Goldeneye and the lovely Goosander. Sometimes too, the alien Ruddy Duck occurs there. This is a native of the Caribbean islands which was introduced to the Wildfowl Trust at Slimbridge in Gloucester from whence it escaped and rapidly colonised southern and central England. In early autumn, the sewage farm was alive with a myriad of flitting Pied and Yellow Wagtails which had bred locally and had brought their 'adolescent' young families to the source of the most insects on which to feed them up, in the case of the Yellow Wagtails, before their hazardous journeyings to North Africa for the winter. This species formerly abounded as a breeding bird along the Lark marshes, but most of these have since been 'improved' by ploughing, and if reseeded as grassland, are still unsuitable for breeding as the new pastures are level and contain at most, one or two species of grass and little other herbage. Yellow Wagtails like a nice tussocky terrain, and position their nests under an overhang of rough vegetation. I was lucky enough to find one on one of the unimproved meadows, with its clutch of five khaki coloured eggs. Since the destruction of the vast majority of the birds' preferred habitat, it has become rather scarce as a breeding species, but in recent years in North Norfolk I have been interested to see adults carrying food obviously to young in nests, in fields of vining peas and potatoes. Before leaving the sewage farm, it is worth mentioning that the wider baulks, the ridges between the plots, were always planted with Bat Willows. To make good cricket bats, they have to grow as quickly and evenly as possible to give the right 'grain' to the wood. On one occasion the farm foreman told me that, due to the moisture and the nature of the effluvia pumped onto the plots, they were able to harvest them as mature trees on an eleven year cycle compared to the normal 21-25 years from sapling to maturity.

Bury became an overspill town after the war, with a rapidly increasing population and much new industry. The sewage farm could not cope with the rapid increase in effluent so it was closed following the building of a modern plant at Fornham All Saints, only three miles from the town. The old sewage farm was allowed to dry out, after which, for several years, it became a council rubbish tip. When capacity was reached, the rubbish mountain was covered with topsoil from building sites and, after compaction, I remember seeing a thin crop of barley on it one year. At this time, the Anglo Saxon village had been discovered and was being excavated, the main area was grassed and the whole became the West Stow Country Park and Anglo Saxon village. Within a very few years, it was being hailed as 'traditional Breckland landscape'! There is still one remnant of traditional Breckland landscape at West Stow, and that is the old pine belt bordering the road. In more peaceful days, this held nests of breeding Redstarts and Crossbill with those of Tree Pipit and Woodlark on the ground nearby, the belt providing the latter two with song perches.

What about the area on the other side of the road with its smothering ranks of pine? What was that like fifty years ago? From my diary for June 12th 1948, three days after we moved to Pear Tree. We set out to catch the bus to go to Bury, but we had not got an accurate timepiece at that time, so we missed it by about a hundred yards. It was a warm, sunny afternoon, so we walked on to Lackford bridge and took the old tow path along the south side of the river. Looking at the trickle that flows down there today, it is hard to visualise the river as it must have been late in the 19th century when horse drawn barges laden with coal used to be taken from Brandon Creek at the junction of the Lark with the Great Ouse right the way up the Lark to Bury St.Edmunds, though the evidence of the former locks here and there along the river confirm this.

As we made our way upstream, considerable numbers of Sedge Warblers and Yellow Wagtails delighted us all along the river meadows. Moorhens abounded, and twice, we disturbed a fishing Heron, on one occasion coming round a bend almost on top of it before it saw us. The water was crystal clear, and we gazed in amazement at an enormous eel wriggling its way downstream quite openly. Fortunate for the eel that we had frightened the Heron! Passing one of the defunct locks, a splash at the edge of the reeds almost at our feet was followed by the sight of a peculiar object swimming away diagonally across the river. It took us a moment or two to decide what it really was as it paddled furiously away under the water, getting its feet entangled in the weed every little way. We agreed that it was probably the most unusual view of a startled Little Grebe that we were ever likely to have. Many noisy Redshank were seen, circling and displaying on quivering wings as they rose and fell using their evocative "tu-tu-tu-tu—-" note ending with the rapid more usual 'song' as they glided down to land on a tussock, or screaming abuse at passing pedestrians along the river bank as they flew across the river for a lunch snack on the sewage beds. Two unidentified duck got up from the latter and flew off. A few days later, we saw them again and realised that they were Gadwall. Later, just having forded a detour stream by another lock, we had close up views of a pair of displaying Kingfishers, so engrossed in each other that they landed on a handrail not fifteen yards from where we were standing, before chasing round and round again, close to us , landing again, and finally flying off having been utterly oblivious of our presence. Having crossed a footbridge over the lock, we made our way across the West Stow road and came home over the heaths with Woodlarks and Stone Curlews much in evidence. We also saw Wheatears and both species of Spotted Woodpeckers in the various belts, some of which were of oak. But pine forests? A nightmare still in the future!

The same evening at dusk, I went alone onto the derelict area across the Icknield Way, and had walked no more than fifty yards when a largish, hawk like bird with stiff winged flapping flight, came silently

View from windows of Pear Tree Cottage, August 19th 1949

towards me, passing within a few yards. After a few moments, it returned and actually circled me curiously within a few feet, revealing the white spots on wings and tail which proved it to be a male Nightjar. We came to know these strange, nocturnal birds quite well during the next two summers, a male coming quite frequently to perch horizontally along a branch of one of the seedy spruce near the bedroom window, where it would indulge itself in lengthy spells of eery churring the sound appearing to rise and fall as it turned its head this way and that. I once found a nest only a hundred yards or so from the bungalow, though it was not in use at the time. Just a scrape in the sandy soil beside an elder which held the two elongated mottled eggs, one of which had hatched, the other being addled.

The day following, a Sunday, we went for a long walk across the heaths to the north, where heat haze shimmered and distorted any stunted elders into grotesque, writhing shapes. This included Weatherhill Heath, about two miles up the track from Pear Tree, since completely inundated with pines and spruce. In the heat of the day,

View from windows of Pear Tree Cottage, two months later

the birds of Breckland move about as little as possible. Stone Curlew, for instance, are crepuscular, and only start to move about much as the sun dips toward the western horizon. One or two were disturbed, Lapwing were driven frantic by our passing, and the many Wheatears 'tucc-tucced' loudly as they flitted from one rabbit burrow to the next in annoyance. Marauding Crows, looking for dead or sickly rabbits seemed to be the only birds on business. The heath appeared to be completely devoid of any vegetation other than a few sparse lichens, the reason not being very difficult to discover, as we had never seen so many rabbits together before. As we walked across towards Davis' Belt, the heat haze accentuated the illusion that a huge grey carpet was rolling up in front of us. Most of the rabbits of Breckland heaths seem a slightly flatter grey than their kindred in other places, no doubt to camouflage them amongst the lichens which are often the most frequent members of the Breck heath plant communities. Among the thousands seen that day were one or two black ones, and, just occasionally we have seen the odd pale sandy coloured specimen.

A few years back, we strolled along a ride which was probably fifty yards wide-much wider than usual-and looked out over Berner's Heath beyond, which still belongs to the Elveden Estate. Very few rabbits, no Stone Curlew or Wheatears, and the grass too long and rank to attract either. The wide ride was across Weatherhill Heath near where we had walked in 1948. Was there no redeeming feature in the monotonous sameness of the pine forest? We had startled a pair of Fallow Deer near the start of our walk and, from Davis' belt, we watched a line of five magnificent Red Deer stags led by a 24 pointer, as they threaded there way down beside another belt half a mile away. The pines that cover what was West Stow Heath when first we knew it, give cover to numbers of Roe Deer and the alien Muntjac, the latter in Kings Forest being the basis for a long term study of the species by Mrs Norma Chapman and a number of helpers. This is necessary as Muntjac are secretive little beasts, lurking in the middle of bramble clumps and producing a higher number of young than the native Roe. They are said to be reaching plague proportions in some areas, though their preferred food seems to be the young shoots of bramble. They are also supposed to do quite a bit of damage to a potentially threatened wild flower, the Bluebell in woods where they are numerous. They are also very fond of sweet chestnuts. We were sitting having a picnic lunch in the car during December in 1998 near the edge of the Stanford training area in Norfolk, when we became aware of a Muntjac about thirty yards away, rummaging amongst the fallen leaves. After a few minutes, it looked up and saw us watching it. It raised its tail to the vertical to show off the white underside, warning any other Muntjac that there were enemies nearby, and disappeared at speed amongst the trees but was back rummaging within five minutes. When we had finished, I was curious to see what the attraction might be, and found emptied chestnut husks and chewed up pieces of chestnut which it had dropped.

Yes, there are now four species of deer in that immediate vicinity, so there have been some gains as well as some losses, and, with areas of clearfell becoming available, the drastic fall in the numbers of Woodlark has not only halted, but numbers have started to recover in quite spectacular fashion in recent years. Most other declines have been maintained. I have no recent figures, but in 1990 the estimated numbers of the native Red Squirrel in Breckland was less than 10, whilst forty years before, it was frequent in every pine belt. In that year also, game wardens in Stanford Training area killed more than 800 Grey Squirrels. A pretty animal to some, but with revolting habits which justly earn it the name of 'tree rat'. In spite of this area being regularly scoured to keep the numbers down, there does not seem to be much difference in the population, and it is even possible that the culling serves to improve the health of those remaining,through lack of competition.

The chronicle of change in the Lark valley since we have known it has been mirrored in almost every other river valley throughout the eastern counties. Many have become arable, and those that have been reseeded and retained as grassland are only sterile copies of what went before. Nowhere for Snipe, Redshank, Meadow Pipits and Yellow Wagtails to nest and the adjacent dykes all too frequently polluted by nitrogen run off and herbicide use to keep the grass a one species monoculture.

The Major's fields by the river that grew that grassy wheat after they were ploughed the first time, were little more than grass stubbles when I came to plough them after that first harvest, with the brand new Fordson Major with its hydraulic mounted two furrow plough. He had to fork out to buy a new one on this occasion as they were an innovation, and second hand ones did not exist . There was also a new cultivator to go with the tractor as well as a number of 'new' towing implements bought at farm sales, some of which would have been better, left where they were. Not that they were a particular hazard on the riverside fields, though there were plenty of other hazards. Whilst meadows, some of them had been quite small parcels of land intersected by drainage dykes. Over the centuries, these had silted up and, when they were first broken up, I suspect that the work had been done by the War Ag. using a crawler tractor. Ploughing them with the new Fordson Major for the first time, I quickly discovered a number of 'no go' areas. The soil was very friable and an easily worked peat, so I would be bowling merrily across the stubble when all at once the tractor would drop vertically for about two feet and everything would stop, and I had landed in yet another silted up dyke. Had I had a towing implement on the back which could have been unhitched and dragged away, things might have been easier, but with a hydraulic plough which was to all intents and purposes part of the tractor itself, I was well and truly stuck. Many hours were spent that autumn, digging round whichever part of the tractor had sunk in. In many cases, the further I dug, the further the tractor sunk, or keeled over sideways at a crazy angle, if it happened to be just one wheel that had dropped in. I hate to think how many times the Major had to call the local farm manager of the Elveden estate to send his crawler tractor and pull me out that autumn, after hours of fruitless digging.

Even so, there were plenty of bonuses if I kept my eyes open. On September 17th, 1948, I was ploughing the field nearest Lackford Bridge, one with no silted drains, so I was able to look round a bit more. I noted in my diary that a flock of Rooks flew over which were the first of the autumn. At that time, the nearest Rookery to Icklingham was several miles away and Rooks themselves were a rarity during the summer. During the following winter, however, we had a roost estimated to be 30,000 strong about a mile from the bungalow in very young trees at the back of West Stow village, roosting on the ground, of course. Crows used to roost in a dense old pine wood which had been planted many years before by the Cadogan Estate. The resident Carrion Crows at that time, had their numbers augmented in October by arrivals of Hooded Crows from the continent. These birds, now extremely rare, were then a regular feature round the sewage farm, though this was not their only feeding place as roosting numbers topped the 100 mark. During the 1960s and early 70s there were still small numbers coming to the north Norfolk coast in the Wells next the Sea area, but these too have disappeared, though no one seems to have suggested a reason for this.

After the Rooks had gone over, I noticed considerable numbers of Swallows and Martins hawking flies overhead, and especially over the river itself.

Suddenly a sickle winged bird came zipping over the bridge right through the middle of the hirundines, scattering them in all directions without actually making contact with one. Looking like a giant long tailed Swift, it continued on its way downriver, becoming ever more hawklike with each wing beat-my first Hobby.

On December 23rd of the same year, I was driving the Fordson along the side of one of the other river fields, ploughing the headland as dusk approached. The light was still very good as it was clear and just about sunset. I looked up to see a large bird arrive with a quick flurry of wings on the branch of a poplar tree about 30 yards ahead of me. A second, slightly smaller bird, circled briefly before flying on. Both birds were slaty blue grey above and white below with black streaking. The one which landed showed feathered upper legs which were pale yellow below the hock, and both birds had heavy black moustachial stripes down from the eye. A pair of Peregrine Falcons! The larger female having landed, flew on to another tree further down the side of the field as I approached, but on the second occasion, she allowed me to stop right underneath her and have a good look without taking alarm, before flying a short distance to a third tree. This time, I went past her without stopping, and had actually nearly got round the field again before she flew off toward Lackford. I noted that the size and flight agreed exactly with a bird seen on December 17th, and also with other mysterious hawks seen on earlier occasions. During the next few winters, peregrines were seen quite frequently during the winter, flying up into the forestry to roost, and the much smaller Merlins were also seen from time to time.

Then followed the decade of destruction for these beautiful birds, caused by the use of the mercurial seed dressings, Aldrin, Dieldrin and Heptachlor. These decimated the numbers of birds at the top of the food chain by poisoning of their prey species. Rats, mice, voles, sparrows and pigeons ate the poisoned grain and were in turn eaten by Kestrels, Owls, Sparrowhawks, Peregrines and other birds of prey. Though many were not directly killed by eating poisoned prey, the pesticide residues built up in their tissues, and while some died a lingering death,most were rendered either infertile or only partially fertile, and subject to laying very thin shelled eggs which broke during incubation. After the pesticides were finally banned, the damage was such that it took most species many years to build up their numbers to former levels. There has been a more or less full recovery, but, paradoxically species such as Barn Owls and Kestrels now suffer problems caused by habitat destruction, there now being very little rough ground for the voles and mice to live among, except, in some areas, alongside motorways and busy main roads where the birds can very quickly become road casualties. Sparrowhawks, on the other hand, are now as common as they ever were, as many people will know who regularly feed small birds in their gardens. Not only will they regularly patrol a garden where birds are fed, they will also find a quiet perch from where they can watch, and from which they will launch a sudden attack on a group of feeding tits or finches unaware of their danger.

On July 11th 1948, I got up at 6.25 and walked down the gravel pit to the farm to milk the two cows and feed the pigs. Willow Warblers and Common and Lesser Whitethroats were singing merrily round the old clay pit at the bottom of the garden. Clouds of Sand Martins wheeled to and fro overhead as I scrambled down the gravel pit, disturbing a Green Sandpiper from beside the pool as I did so. Having finished the stock work, I made my way home by a circuitous route which took me along the edge of the forestry land and brought me back to the Icknield Way above the cottage. In doing so, I found a colony of Lesser Meadow Rue in a chalky spot, a plant which we found quite frequently, but which seems to have declined drastically since then, and many of todays sites are obvious garden throwouts. A Woodlark was disturbed, propelling itself skywards like a big butterfly until it had gained sufficient height to drift round and round on flickering wings in the still air , delivering its beautiful song. A male Red-backed Shrike watched me curiously from a distance of about seven yards as I made my way slowly past. It undoubtedly had a nest very close by. Sad to realise that this bird no longer breeds regularly in the British Isles . The last breeding that I know of was in a traditional site not ten miles from Pear Tree Cottage in 1989, though there have been a few reports of odd pairs breeding or trying to breed at undisclosed sites in southern Britain since then. In 1948 and 1949, they could be found commonly throughout Breckland and down the Suffolk Sandlings coast, and in the latter year. there were actually three pairs within a quarter mile of the cottage, two of which successfully reared broods. After breakfast that day, with a few hours to spare before the afternoon stock work, we set off on our bikes, through Icklingham and onto the Eriswell road, with Corn Buntings singing from the wires every little way. These too are much declined, though there is still a small population along the Fen edge with Breckland. One is tempted to wonder if their decline is anything to do with the putting underground of all the electricity and telephone wires as the cornfields in which they frequently nest are still everywhere to be found, and in an area such as the Brecks and Fens, the fields have altered very little in size and shape. The theory that the decline of this and other seed eating species such as Yellowhammers is to do with the early ploughing of stubbles after harvest, resulting in a dearth of weed seeds for the autumn flocks of young birds to feed on is probably the correct one.

The suggestion that the decline of the Red-backed Shrike is due to climatic changes does not really hold water if we are to give credence to global warming, as this is a bird of more southern distribution at the northern end of its range, so one would have expected it to be expanding rather than contracting its range.

We went on from Eriswell to Wangford and Brandon where we followed the Bury road until we reached Mayday Farm. Here we turned into the old drove known on some maps as the 'Shakers Road', still a public bridleway through to Elveden, and certainly in use for many hundreds of years. Today, Mayday Farm is the start of Forest Enterprise's Goshawk Trail. Here, it is claimed, it is possible to see and hear breeding Woodlarks, Nightjars and Goshawks in their traditional surroundings. This is an area where some of the clearfell is not being replanted, giving a permanent clearing which is being managed in such a way that the first two species will still come there each summer. Only the very fortunate ever see a Goshawk as the trail only passes through a corner of their hunting territory, and very sensibly stays well clear of the actual breeding sites, which vary very little from year to year. For instance, a nest built in 1981 was not used in that year. However, it was used, and probably by the pair that built it, for the first time in 1991! The Goshawk is a relative newcomer to the forest, and it may be conjectured that some at least of the first two breeding pairs had escaped from captivity. Nevertheless, their young are wild hatched, and they have now become well established in the forest with 8 pairs present in 1997 from a national total of around 300 pairs. We have been to Mayday on a number of occasions since the trail was set up, and on just one occasion we were fortunate enough to see one of the birds drifting majestically across the clearing after sunset when we were waiting for Nightjars.

Today, reaching the southern margin of the forest, where the Shakers Road emerges onto the eastern end of Lakenheath Warren, it is difficult to decide where forest ends and 'warren' begins. Over the years since myxomatosis decimated the rabbit population, the open heath has become over taken, first by coarse grasses, and subsequently by dozens of seedling Corsican pines spreading inexorably out from the forest, and now almost halfway across the Warren at that point. On our first visit, things were very much otherwise. Soon after leaving Mayday Farm, we stopped to eat our sandwiches, by which time the lovely sunny morning had given way to scudding clouds. Setting out towards Elveden as the first drops of rain came splashing down, the scene changed rapidly. We followed through a cathedral of black dripping trees while the rain became a deluge. We emerged abruptly onto the Warren, where soggy rabbits scuttled for cover, not into their burrows but behind stacks of bombs which stretched for as far as the eye could see across the heath. Admittedly, there is something of a depression at that point, so visibility was probably about one hundred and fifty yards. Wheatears perched on six feet high stacks of hundred pounders, five hundred pounders, perhaps even 'tonners', to scold our intrusion. A couple of Stone Curlew eyed us warily, ran a yard or two and lifted off, revealing the lovely black and white wing pattern in their otherwise crytic desert coloured plumage.

We were very much taken aback, but there was no notice to say that we could not cross the bomb dump by the traditional route, and I had to get back to milk! I wonder when they cleared them up? It was already more than three years after the end of the war. The cold war had not started and I would imagine that the Lakenheath airbase was in some kind of limbo. The only planes we ever saw or heard were occasional aircraft taking off or landing at Mildenhall, five miles away, but that had been a permanent aerodrome since well before the war. In fact, it was the take off point for the England to Australia air race in 1932, and I remember watching eagerly that morning to see if I could see any of the competitors as they set out on their long journey. I was not completely disappointed, as three did come within earshot, though only as specks well down on the horizon to the west of Hitcham. After the beginning of the cold war, both airfields took on a new importance, with the Americans using Lakenheath for their most sophisticated warplanes, while Mildenhall became one of their most important transport bases in western Europe, with giant freighters going to and fro across the Atlantic at all hours of the day and night. From time to time during the cold war, some of their high speed, high altitude reconnaissance aircraft such as the Blackbird paused to refuel at Mildenhall. With the disintegration of the Soviet empire, one wondered what would happen to bases such as these which for so many years had played a vital role in the economy of the surrounding district. Today, and for many years past, a trip to that part of Breckland can be an earshattering experience, especially near Lakenheath which was equipped for many years with Fl-lls, supersonic fighter bombers that were used to strike at Col.Gaddafi in Libya, flying directly from Lakenheath and using mid -air refuelling to complete the long journey. At the time of writing, with continuing troubles in the Middle East with Saddam Hussein in Iraq, and nearer at home in the former Yugoslav republic, there seems little chance that the Americans will withdraw from their local bases. The Fl-lls have been replaced by F16s but their routine is still the same. Some days they will be practising take-offs and landings seemingly endlessly in pairs, the noise of their engines echoing through the forest and making ordinary conversation impossible.

March 21st 1949. At seven p.m. I cycled up the track to Weatherhill Heath in the hope of hearing an early Stone Curlew call. As I was approaching the heath, I heard the unmistakeable call, rendered locally by some as 'Willie Reeve-willie reeve-willie-reeve." giving the bird one of its Breckland nicknames. It is also known as the 'Norfolk Plover' or 'Thickknee'. Willie Reeve would seem more evocative.

Reaching the edge of the heath, I waited in the gathering dusk for more than a quarter of an hour without hearing it again. However, there came to me a continual multitude of sound from far and near, of literally dozens of Lapwing tumbling about in the darkening sky uttering their nuptual calls

"Weep-weep-weep,pesiaweep-weep-weep—", the only other sound, the throbbing of their wings on the still air as they cavorted wildly this way and that with continual changes of direction. Of all the tumbling birds, I failed to see a single one in the gathering dusk. Had the F16s been going down to Lakenheath or a Galaxy into Mildenhall, I would not have heard one either.

A species recovery programme for the Stone Curlew has been gathering momentum in recent years. In all the Brecks of Norfolk and Suffolk, 87 pairs were located during the summer of 1990 (Suffolk Birds 1991). This represented a small improvement over the few immediately previous years. 84 of those pairs bred or tried to. The Norfolk Bird Report for the same year records that 37 pairs fledged 29 young in the Norfolk part of Breckland. Assuming the same success rate for the 47 Suffolk pairs, which would suggest 40 fledged young, a total autumn population in that year would have been 150-155 birds. The 1997 figures issued by the Royal Society for the Protection of Birds who are in charge of the species recovery programme reveal a total of 122 pairs in the whole of Breckland which reared 114 young, giving an autumn total of 236 birds. This suggests that the programme is working, but that there is no room for complacency when one considers that not only are the RSPB involved, but also the many landowners on whose land the birds nest. Very many of the nesting pairs now nest on arable, and especially amongst such crops as carrots, so all the farm staff who have to do with weed control and other farming operations amongst such crops as the birds are nesting in must be really genned up on what is expected of them to make the recovery programme a success. In 1948 and 1949, we would have ridiculed anybody who had suggested to us that within a very few years the Stone Curlew would become an endangered species. During the late August of 1948 we were cycling up the Icknield Way towards Shelterhouse Corner when we came upon an area where the young pines had partially failed, with just one alive here and there. Suddenly a vast pack of Stone Curlew took to the air and started to make their way back towards Wordwell where there were still large areas of unplanted heath. We were not able to do an accurate count, but we know that there were more than 200, and they were back in the same place when we went that way again a week or two later. We also knew of another pack in excess of fifty birds on Tuddenham Heath at the same time less than two miles away. During September 1949, I cycled from Icklingham across Tuddenham Heath, thence towards Cavenham, returning by way of the Cavenham to Icklingham road, no more than five miles in all. My diary says that parties of Stone Curlew in ones and twos to thirty or forty were getting up at frequent intervals throughout my ride. Even when one considers the latest figures, I would suspect that the current population can be little more than 10% of what it had been fifty years ago.

Breckland was not all birds however. Rabbits have been mentioned, Hedgehogs and lesser things like Stoats, Weasels, Voles and Shrews all came our way, and we did see the very occasional Roe Deer, though at that time, these were mainly confined to the Cavenham/Tuddenham Heath area. Late one evening by the poplar plantation at the end of the Major's land, we witnessed a gathering of 250-300 medium sized bats beating to and fro round the plantation and especially low over the river. With hindsight, I suspect that these could have been Serotines.

On June l9th 1949 at 9.00p.m. we walked up beyond the top of the first wood. There were two smallish oak woods, relics of the Cadogan estate, in the first mile of forest above Pear Tree Cottage. We were looking for Nightjars on this occasion and stationed ourselves at the top end of a wide ride through a mixed larch and pine plantation. A male Nightjar was seen briefly and we heard him wing clapping somewhere nearby, part of the display flight. As we waited, a Hare appeared in the ride about a hundred yards away, coming towards us. We stood motionless watching, while it paused to feed a little. Presently it came trotting towards us again, stopping at about 40 yards, totally unaware of our presence, though to the hare, we were outlined against the evening sky. Now it sat up on its haunches and proceeded to wash its face using both forepaws at once rather than one at a time as a cat would. Next, it washed and groomed some of its underparts before resuming its supper. After a while, it came on again, actually halving the distance between us, and at such close quarters, looking remarkably like a young donkey from the front. Walking in a peculiar fashion, and not without a certain grace, from one dainty morsel to the next, we saw it snip off a stem of cocksfoot grass with a single bite, eating the most succulent portion before letting the remainder fall to the ground. Every detail of this beautiful creature could be seen as it fed its way slowly past. Supper over, it trotted contentedly along the top margin of the wood. I measured the distance from where we stood to the nearest point that the hare had approached us. Fourteen yards. One of those magical occurrences that only come once in a lifetime.

Life was not all idyllic, even in the natural world in which we found ourselves. June 25th 1949. During the early hours a pesky Nightjar came into the tree by the bedroom window and churred for what seemed like an endless age, keeping us awake. As bad as the Nightingale earlier in the year!

At this time, we were still learning the plants of Breckland, our only aid 'The Illustrations to the British Flora' illustrated by W.G.Fitch and companion work to that old classic, 'Bentham and Hooker'. We did get this latter work when we could afford it, but at first, we just had the 'Illustrations' which were all line drawings, four to a page and the pages seven inches by four and a half. It was surprising how much we were able to identify by looking all through the book when we found something we did not recognise, which says a great deal about the accuracy of the drawings.

On September 22nd 1948 we walked across Mr Wagstaff's barley stubble on our way home from the sewage farm and came upon a small flower of the deadnettle family. We were surprised as, apart from Yellow Archangel, a woodland plant, we did not know of anything in that family with yellow flowers. Looking through that family in our illustrations we soon discovered Ground Pine and happily added it to our growing list of new plants. When we finally got our copy of Bentham and Hooker, we discovered that Ground Pine was said to be very rare. It couldn't be! In Mr Wagstaffs barley stubble there had been more than a quarter of an acre yellow with it. We visited it each year after that to make sure that it was still there. Some years later, Francis Simpson started collecting records for his Flora of Suffolk and appealed for help. I sent him some lists of things we had found in Breckland and elsewhere and he wrote back demanding a specimen of Ground Pine, for which he had no other records. I was able to send him one to prove it. Later, Miss J.C.N.Willis took over collection of records for a while, came across our record and also required verification and again I was able to send a piece. Next followed the collection of data for the first Atlas of the British Flora organised by the Botanical Society of the British Isles, and Dr Frank Perring also had to be put in the picture. Somebody else visited the field and spotted the plant and claimed it as a new discovery. They were, however, quickly disillusioned. It must have been about 1985 when I last saw a a plant-or two and it was, oddly, in December when the field had been left stubble for rather longer than usual. Unfortunately the field now belongs to an absentee landowner said to be unsympathetic to conservation, who has grown intensive carrots on the field a number of times and it is feared that the plant may now be extinct there.

Not the end of story! In 1989, Mrs Enid Hyde, the then BSBI recorder for Suffolk, wrote and told me that while going through some papers belonging to the late W.H.Mills, a botanist of international repute, she had come across a reference to "Ajuga chamaepitys (Ground Pine.) In a field at West Stow to the east of, and adjacent to the Icknield Way August 1937." So WHM found it eleven years before we did, but did not report it at the time.

That was one of the rare plants found in Breckland in 1948 and 49. It was a special record though, as it was not one of the several very rare species for which Breckland is famous, being at the extreme north of its range. Its other British sites are in a few chalky fields south of the Thames, and when one gets as far south as central Spain, it is a common constituent of the waste land weed flora, especially where there is chalk or limestone in the soil.

The special plants of the Brecks are species of steppe habitat, and only occur in arid conditions on poor soils that are sunbaked in summer and very cold in winter. The shallow basin which is Breckland has the right type of soils left behind by the last glaciation. Because of this, it is not uncommon in sheltered spots, to find bracken shoots which have been recently frosted in any or every month of the summer. These are soils which can be worked at any time of year as soon as the rain stops, and which encouraged the Neolithic farmers to break them after clearing what must have been minimal tree cover, probably mainly birch. After farming them for a few years, until all the nutrients were exhausted, they would abandon each plot and move on to another. In this manner arose the farming patterns of the district which remained in place until the second half of the twentieth century, when the upsurge in the use of chemical fertilisers allied to pesticides of all kinds and latterly linked with spray irrigation which now allows farmers to grow top quality crops from blowing sand.

On June 9th 1949, I discovered a plant growing quite near the Lesser Meadow Rue along the edge of the forest, which looked like that plant but had erect rather than drooping florets. Unable to find anything like it near Meadow Rue in the illustrations, I noted it down as 'Upright Meadow Rue'. At that time, the areas covered with pine above Pear Tree, were in a number of discrete blocks, partly damaged by troop manouevres during the war, and with a wide strip of derelict Breck round each block. These strips were probably between fifty and seventy yards wide and were havens of occupation for Woodlarks, Tree Pipits and Stone Curlew amongst a garden of Breckland wild flowers. These included species which some doubt as being native in Britain, such as Star of Bethlehem, which grew there in some quantity, and which would surely never have been introduced into such a remote situation. Its normal habitat is steppeland and steppe type heathland. It is able to withstand disturbance, so it was very much at home there. Growing near some Star of Bethlehem on June 16th that year, I found a strange member of the Campion family with small, bright pink petals protruding from an inflated, green and white striped calyx which was covered with sticky, glandular hairs. Looking through the 'Illustrations', I soon spotted Striated Catchfly, now more usually called Sand Catchfly. Rita, who was looking over my shoulder, pointed excitedly at the next illustration. There was our 'Upright Meadow Rue', only its proper name was Spanish Catchfly. It was not until much later that we discovered that both are 'Breck rarities' which at the present time are 'Red Data Book species'.

Maiden Pink formerly abounded on the Breck strips round the plantations, and all up the side of the Icknield Way, sometimes large patches of it. It became much scarcer as all the unplanted areas were infilled and, after myxomatosis which struck Breckland about 1955, was choked out from many places by coarse grasses within the space of a few years.

A few days before the above, we had had a week's holiday, visiting relatives, returning home on June 6th. Late in the day, after visiting our various nestboxes and filling the 'birdbath', we were greeted at dusk by our local Nightjar who came and sang from his usual perch near the bedroom window. My notes tell me that I was then able to contradict the 'Handbook of British Birds' and other authorities which state as I have already done elsewhere in this book, that the rise and fall in tone of the Nightjar's song is caused by the turning of its head from side to side. We crept unseen to the bedroom window and were able to watch its every movement for several minutes. Although the head may sometimes be turned at the moment of the rise or fall in pitch of the song, this is not the cause of the change of tone. At irregular intervals, we clearly saw the beak, which was fairly wide open for the duration of each burst of song, close momentarily from halfway open to only about a quarter of the way open, thus changing the pitch for a split second. Moreover, the fall and rise effect was produced even when the birds head was motionless other than from the beak movement as well as when it was turned. The following morning, one was churring for two or three minutes in brilliant sunshine at 7.25 a.m., a very rare occurrence in our experience.

By the spring of 1949 we had already become more restricted in our movements as, during the winter, our first daughter, Susan, had arrived. Babies were all delivered at home in those far off and often primitive days. My mother's mother, Grandma Somers, had come to stay a day or two before the expected arrival which, though a day or two early, was still long and arduous. I first cycled down to the village to alert the midwife, the local district nurse, Nurse Crack, at five o'clock in the morning. There were occasional pains throughout the day, though the nurse had had to get on with her rounds, but she looked in again later. Late that evening, she sent me down to phone Dr.Rees from Mildenhall. He came out and made an examination, deciding that nothing was going to happen yet, so he gave Rita some painkillers which looked like the old fashioned Pontefract Cakes and told her to get some sleep. She did not. The 'painkillers' actually seemed to make matters worse. All through the night the nurse stood by and tried to help, but it was not until the early hours of the morning that I was again sent to phone the doctor. He finally arrived at about 7.30 and Susan made her arrival known with a lusty bellow at almost 8.30, weighing in at 10 lbs.

All that time, I had had to keep the copper hot to supply the nurse with hot water. The kettle was boiled over the oilstove umpteen times to make cups of tea. Worst of all from the point of view of my peace of mind, Grandma Somers whiled away the dark hours of the night reminescing on difficult labours she could remember from her young days back in the 19th century!

Rita recovered quite quickly despite her long ordeal, though there was one minor hiccough a couple of days later, when I arrived home from work to find her running a temperature. The doctor had been in and prescribed some tablets which I had to cycle the seven miles to Mildenhall to collect by a certain time. I had no lights for my bike at that time, but fortunately there was a full moon. All went well until I was back to 'our' side of Icklingham, when a car drew up revealing two policemen. Naturally enough they wanted to know who I was and what I was doing, cycling in the dark without lights. I told them my mission and they said I would have to walk the rest of the way but, in those days of little traffic, let me off with a warning and drove on. I got off when I saw any other lights approaching!

Necessarily, after that, we only visited places together that were within walking distance. Rarely, we would have a few hours out while the other babysat. Even that was scarcely necessary as there was so much to see on our doorstep.

I only have one or two day's notes made whilst out alone, during the remaining months of our stay at Pear Tree Cottage. For instance, on July 24th 1949 I surprised myself by waking up at 4.30 a.m. BST, getting out of bed to find a nice morning, and compounding my astonishment by getting dressed and going out into it! A late Nightingale was singing in our lower hedge and was being answered by a second over by the sewage farm. By ten to five, I was cycling down the track towards Lackford Bridge. This was the ricketty 'old' bridge which was bypassed and demolished a few years later. The old bridge was at the crossing point of the original course of the Icknield Way, and in earlier times would have been a ford. The present bridge was built for the main road, the All0l Bury St.Edmunds to Long Sutton road where it links with the A17. Sadly, though there is the remains of the former tarmac going up to the former crossing, this will disappear completely in a few more years. As I crossed the West Stow road near Blyth's(Letch Moor cottages), the pair of summering Whinchats were seen on the heath nearby. Never a common breeding bird locally, it nested on West Stow Heath that summer, and was findable in several other sites where the river marshes and heathlands met.

When I reached the bridge, I found the early morning mist hanging over the river so thickly that I could only identify Mallard with any certainty. Reed and Sedge Warblers were both singing well nearby, as were several other species which I thought had stopped for their mid-summer break. I stopped at the corner of the sewage farm to see if anything was visible on the pool favoured by Green Sandpipers, but could see nothing. However, a Crossbill flew out of the pine belt overhead and switchbacked away chupping loudly.

Leaving my bike, I went down the edge of the Alder carr fringing the sewage farm at the eastern end and went in over the sewage pipe bridge to look round the lower half dozen lagoons-the ones not visible from the foreman's bedroom window-and found good numbers of Green and Common Sandpipers, two Redshank, a Snipe and several Gulls, Lapwing, Pied Wagtails and Moorhens. A quick nip to the nearby riverbank startled a Kingfisher before I made my way back to my bike. I continued my journey through West Stow village, beyond which I turned into Culford Park. That end of Culford Lake runs up to the right of way that crosses the park, and on this occasion I found family parties of Mute Swans, Canada Geese, Mallard, Moorhens and Coot much in evidence. In the trees overhead, a singing Blackcap and a family party of Nuthatches attracted my attention. Suddenly THE bird was singing in the marshy meadows behind me. Rita had already heard it and told me about it. No inhibitions in those days. I hastily dumped my bike out of sight, dived through the fence by the little stream that takes the overflow water from the lake and proceeded with reasonable caution. The bird I sought was a Grasshopper Warbler, which neither of us had seen at that time. Following the stream in the direction of the 'reeling', I soon discovered that the ventriloquil effect of the song made it appear sometimes to be coming from the herbage at my feet, while at others it seemed still some distance away. Suddenly, I realised that I was quite close to my quarry, as I spotted a bird's tail protruding below a twig in a little Sallow bush about fifteen yards away. Cautiously I edged forward another yard or two until I had the songster in full view, and was able to see every detail of his stripey plumage as he sat unconcernedly pouring forth his continuous 'fisherman's reel' note, beak rippling and body quivering slightly in the early morning sunlight. After several minutes watching, I moved slightly and he apparently became aware of me for the first time, flying off to another bush with a jerky, Whitethroat like flight.

Satisfied with a view of a Grasshopper Warbler that has never been bettered, I returned to my bike and went the quarter of a mile past West Stow church down to the River Lark at Flempton lock where I saw 'another' Kingfisher. It could have been the same one, as I was still only half a mile from the sewage farm. Through Flempton and Lackford

Lackford Bridge, August 1949, at the Icknield Way crossing

villages-a family of baby Spotted Flycatchers were being fed in the grounds of Lackford rectory-I went straight on to Weatherhill Farm where I arrived at seven o'clock to milk the cows and feed the pigs, having already seen 63 species of birds that morning. After breakfast, with Susan in the pram, we walked down to Lackford bridge, stopping to admire one of the local Red-backed Shrike males, perched on top of an elder near the track. Beyond the river, near J.D.Sayer's, we turned up the Cavenham road and in another hundred yards, took a track which, until that spring, when they were ploughed up, had led to Clamps and Home Heaths. The track gave excellent views of a stretch of the river which, on that occasion, were packed with duck families, with Teal and Gadwall in the majority. Reaching a little dip, we glanced ahead and noticed several things simultaneously. A Kestrel got up from the track and a second from a nearby fence post. A Red-legged Partridge was behaving as if it was badly injured. Following the Partridge with obvious intent were three Stoats, who had not noticed that they were being led away from a family of half grown young Partridges by the astute parent bird. We could not be quite sure as to the role of the Kestrels. Had they too been watching the young Partridges with a view to nipping in and having one apiece while the parent was occupied? Or were they quietly mobbing the Stoats which are the enemies of all kinds of birds including birds of prey?

Reaching the edge of an area of newly broken heathland, we noticed a Stone Curlew standing quietly, so searched all round thoroughly with our binoculars. Some of the ploughed land in the middle of the new breck was much sandier than the rest, and here we discovered five more Stone Curlew squatting down or standing and preening themselves in the sun, and generally taking no notice of us whatsoever. I did not say at the time, but I would imagine that they must have been at least 150 yards distant to have ignored us completely. We carried on along the track as far as we were able, before turning into a second which led us back to the road somewhat nearer Cavenham. Before we got back to Lackford, we had seen three family parties of Red-backed Shrikes, two of which were well grown. These would have been in addition to the two breeding pairs on our side of the river.

Back home again, a family party of Crossbills and a Lesser Spotted Woodpecker flew over the garden, the latter changing course to land briefly in one of the old hawthorns before flying on again. After afternoon milking and feeding, we ended the day Nightjar hunting, having tremendous views of a churring male perched along the branch of a dead tree. We also disturbed a Little Owl to give a final flourish to my day's list which closed at 74 seen within a two mile radius of the bungalow. This was without trying to see every last bird, and I heard four more species sing which I did not see. I suspect that the maximum possible in that area today would be far fewer, even two months earlier, May being considered the 'best' month by 'day list twitchers'.

What other naturalists were operating in Breckland in those days? certainly far fewer than there are today. For one thing, far fewer people had transport, binoculars were scarce and expensive, and the range of reference books, apart from the five volume 'Handbook of British Birds', was fairly limited. There were other lovers of the heaths and their wildlife, especially young birdwatchers. The late David Butt and Don Dorling were both at Culford School. We often saw David at weekends, cycling up the track, either alone or with one or two others of like minded youngsters. John Denny we saw quite often. He worked at a nursery near Ixworth and spent much of his spare time cycling round Breckland. A few years later, he joined up with Jehovah's Witnesses and disappeared from the Breckland scene rather abruptly. Tony Vine was then in the RAF and stationed at Mildenhall. His father was station master at Shippea Hill, out in the Fens. His mother's family had connections with Bagmore Farm at Stanford, long since demolished by the military as it is near the centre of the Stanford training Area. Tony, Rita and I are all currently members of the STANTA Bird Group, so keep in touch through a connection with a different part of Breckland to that which we then knew. We saw Tony quite often in those Icklingham days as he tended to drop in just before tea and stay until after supper.

One character that we saw from time to time introduced himself to me in a rather curious manner. We had discovered that a pair of Short-eared Owls were summering and presumably nesting among the small trees somewhere in the two miles by one block above the bungalow, during 1948. We saw them on a number of occasions, as this species hunts by day and especially during the evenings, throughout that summer , after which they apparently moved away. During the following spring, John Denny called to tell us that he had watched the pair hunting over West Stow Heath, a few hundred yards away. I spent some time watching on two or three occasions without any luck, and on the third occasion, I was approached by a tall, slim, rather rough looking gentleman of middle years, pushing an old upright bicycle, his trousers nipped in at the ankles with cycle clips. He was clad in a long, dark overcoat and old cloth cap that had seen better days and he was wearing gold rimmed spectacles.

He came up and greeted me with "Anything about?" This is a frequent 'twitcher' greeting nowadays, but in 1949, it was unheard of. I replied non-comittally that I had not seen anything.

"Oh! I wondered if you had seen any Stone Curlew. There are usually some about here."

I again replied in the negative. He looked keenly at me, sensing my reticence. Then he went on in a more animated fashion. "I'm a bloody naturalist, see? I're bin laid up, and I're just biked out from Bury to see all the bloody spring flowers afore they're over."

I learnt that his name was Henry Boreham, and that he worked for Bury St. Edmunds corporation in the sewage department. In addition to being something of a rough diamond, he was also a very good naturalist, spending most of his spare time on the heaths looking not only for birds and plants, but also for insects. Over the years, he carried out a painstaking study of the habits and life cycles of several species of Digger Wasps, solitary wasps who excavate burrows and stock them with titbits including caterpillars, on which their own larvae feed. His work was of sufficient importance and interest for him to be made an Honorary Life Member of the Suffolk Naturalist's Society, and his doings were a byword among other followers of Breckland wildlife

My suspicions of him at our first meeting were not without justification. Not against Boreham, as he was always known, but because Tony Vine had told us that professional egg collectors were operating in the Brecks, and some lived not too far away. We knew their names, but not what they looked like. Nuff said. I do not know if the ringleader is still alive or not. I did hear that he 'reformed' later on.

As the hot, dry summer of 1949 wore on, problems at home started to mount. Not domestic strife or anything like that. From our earliest days we had come to an unspoken agreement that a quarrel was not carried over from one day to the next. Each day was started in pristine condition, and so we have carried on ever since. Certainly, we have had our ups and downs, but with each day new, personal ones have been short lived. This particular problem lay, unfortunately, with the Major. The top of the well which housed the winding gear was rotten, and, try as we might, he prevaricated about getting anything done about it as it meant having a carpenter make a whole new top, which meant spending money. Animals started falling sixty feet into the water and coming up drowned in the well bucket. A mole or two, followed by a rabbit, but the final straw came when I hauled up a dead rat. We had a small baby and the well was our only source of water for everything that particular summer when the rainwater barrels had been used up. He did run us up a couple of buckets of drinking water from the farm, but that was no answer. Even in those days, it was not feasable to cart all the water for a family of three for three quarters of a mile.

I scanned the columns of the 'East Anglian Daily Times' and, the following Saturday spotted an advert for a cowman wanted on a farm at a place called Timworth, only five miles away, for a wage of £6 a week. I cycled over and saw the farmer, whose name was Shropshire and who, I later learned, came from that county. He showed me round, took me to look over two cottages, both of which were empty, then stood and looked me up and down through seemingly half closed eyes.

"So you think you'd like to be a cowman do you?" I replied that I would.

"Reet!" he said-he still spoke with a broad, Shropshire accent-and that was that.

The last job that I did for the Major after I had worked my notice out, was to go down to Icklingham to the carpenters' to collect a new top for the well. Rather a bitter pill to swallow, as we both hated having to leave our first home, and now realise more than ever what a marvellous place it was for wildlife, though only a shadow of its former, 1940s self.

Never mind, I really was properly connected with the lowing herd for the first time.

Timworth

"Where azackly wuz you then?" Jack fixed me with a steady stare from his bright blue glass eye, his good eye almost hidden by wrinkles and a bushy eyebrow as he regarded me curiously. I explained that you went over the bridge on the Mildenhall road and turned right. After half a mile you turned up a track on the left by a pair of flint cottages. Realisation dawned.

"Oh ah blast yis! I know where y'are now. Yew lived up the side o'th'ole Ickleway Road. I know th'ole Ickleway Road. All them trees what the Forestry ha' planted this past fifteen yare, I 'member them when they wuz all filds. When old Lord wuz alive, me an' me mate, we ploughed all them filds up th'ole Ickleway Road."

Old Lord, the Marquis of Cadogan, in addition to owning the estate which now embraces Kings Forest, also gave his name to the pub in the next village of Ingham, though, during our time at Timworth, I never heard it called by any other than its original name of 'The Griffin'.

At the time we moved to Timworth, it still had the insularity that is peculiar to fairly remote communities, even though it was only four miles from Bury St.Edmunds and on a fairly good bus route. Thus, newcomers had to give an account of themselves to those already in place as local residents.

Jack Plume was horseman at Green Farm, and thus nominally, the most important worker. Though he still had five or six horses in his care, with which quite a lot of the landwork was done, the horseman's craft of ploughing was, for him restricted to laying all the tops and opening all the furrows ahead of or behind the two tractor drivers, who did all the bits in between. How long the Plumes had been in the village, I am not sure if I ever knew, but they had had three sons and three daughters, the elder of the latter being named 'Mons' which rather gave away her age. I suspect that Jack may have been in that battle. The eldest of the sons, John, who was gardener/groom on the farm, the Shropshires keeping two or three riding horses, had been in the Royal Army Veterinary Corps during the second world war, looking after and helping to treat mules injured during the battle of Monte Casino in Italy, amongst other postings, where he was attached to the Eighth Army. Until then I had not realised that we had had animals on active service apparently right through the war.

Jack and John and his next elder brother were all thickset and stocky, with weathertanned complexions and the appearance very much of French peasant stock. To heighten the illusion, Jack had a full, bushy moustache and John favoured a black beret for work. The latter told me that family tradition had it that their ancestors had come to England with the Hugenots some 300 years ago.

Gerv(Gervaise) Shropshire, whose family originated from that county, was a big, florid man, slow of speech and seemingly easy going, but who could flare out in sudden unexpected bursts of temper which often had far reaching effects, as I was to find out in a couple of years time, and so did Jack Plume at some point before then. Jack left, and went as horseman to the only other farm in the village. He did not have to move as he still had three sons working at Green Farm, and Gerv still had a cottage empty, the one we did not take when we moved there. The empty cottage was the middle one of three, built, as were most of the blocks of three on the Ampton Estate, in Victorian times. From the front, they appeared to be pairs, with a front door at either end. Go round to the back and it was to find three back doors so close together as to be almost on top of each other, and the middle cottage with only the one entrance. Gerv managed to find a horseman and his wife willing to live in such cramped conditions, and one who failed to gain any degree of popularity on the farm in the time that I knew him. Probably about fifty, he had the fawning manner of the classic 'creep-arse', as such were called in Suffolk, and a seemingly inexhaustable store of tall stories which he always prefaced with "Let me tell yew. An' what I'm a' goin' to say is as trew as I'm a standin' here in front o' yew, without a wud of a lie—-" and off he would ramble for the next twenty minutes into some of his exploits when he had worked 'in the stoon mines up the sheers'. These were the limestone quarries in the area of Jurassic limestone in what was east Northants , Rutland and south Lincs. Or when he had walked down to Hereford or Worcester to help with the hop or fruit picking. He was an evil man who would lay into a horse with far greater force and brutality than any slight misdemeanour warranted. He also had one eye, lost in an accident in the 'stone mines'. Instead of a glass eye, he sometimes wore a black patch, but more often faced the world with an angry red socket.

We lived beyond the bottom of the Green, in half of what had, in bygone days, been the 'Royal Oak' public house, but which had long since been divided into two dwellings. The population of the village in 1949-51 was between 90 and 100. The Griffin at Ingham was only a mile away, so the early loss of the pub in Timworth was understandable in such a small community. Our next door neighbours were Fred Booty and his family. A daughter who lived in Bury kept a greengrocers shop which is still there in Northgate Street, while her brothers and sister

No.6, Timworth Green, formerly the 'Royal Oak'

had several milk rounds in the district, helped by their parents. Perhaps that is badly put, as old Mrs Booty ruled the family with a rod of iron!

The milk was sold, either bottled, or direct from the churn. Caroline, the daughter living at home, had a roomy, but even then ancient van, with which she delivered milk and greengrocery to villages in the district. She also had at least one daily rendezvous with her brother Jack, who served Fornham St.Martin and part of Bury itself, selling milk direct from a gleaming brass churn which was firmly fixed to the floor of the milk float pulled by a strong bay cob. The milk was served into the customer's containers using pint or half pint dippers. Carrie's rendezvous was to top up the milk in the churn, done in the open air in whatever the weather. Looking at the regulations governing the production and sale of milk in 1999, one is astounded at what 'progress' has been made in half a century.

Jack Booty was the third person within the population of Timworth with only one eye. Like Jack Plume, he wore a glass-eye, and he also lost his during the first world war. Now there is a point for the statisticians to ponder. Three per cent of the village had only got one eye, which would have been six of seven per cent of the male population!

Jack Booty and his youngest son Percy, in addition to helping behind the scenes with the milk business, also had several small plots of land between Timworth and Culford, where a third son lived, all of which they farmed using only horse power.

At Green Farm, the horses were kept at work, though the tractor drivers had plenty to do as well. The binder was a tractor drawn McCormick 7 or 8 feet cut driven by the power take off. When it came to carting the sheaves, all the work was done by horse power.

Moreover, coming from the west Midlands, Gerve did not seem to have much enthusiasm for large wagons which were a picturesque part of East Anglian farming- the ships of the fields, I often thought, with their graceful, curving lines. Instead, all of his off field carting whether it was of hay or sheaves, was done with tumbrils using large fore and aft ladders. Admittedly these held quite a good sized load if loaded properly. At that time, most of the calving was done in October and November to take advantage of higher milk prices during the winter months. This meant that at harvest time, the cowman was slack and could give a hand in the harvest field. I soon learnt that tumbrils with harvest ladders needed careful loading as, being only two wheeled, if too much weight were loaded onto the back of the cart the load would 'hang' and the poor horse between the shafts was likely to be lifted off its feet by the girth strap. Gerv had an old fashioned way of addressing his men. The single workers were all known by their christian names, but as soon as they married, or if they came onto the farm already married, they were referred to by their surnames only. This could lead to confusion if referred to indirectly, as the tractor drivers were married brothers Ted and Lenny Ashford, the latter being married to one of Jack Plumes daughters. Also on the farm were Charley Butcher and Ted Sayer and his son Maurice. Charley and Ted were both old soldiers, and Charley still wore hobnailed boots and puttees all the year round. Ted, on the other hand had seen service in 'Messerpertamia' and suffered from malaria from time to time. The Sayers and Ashfords lived along the bottom of the Green, at the corner of which dwelt Ernie Palfrey, a long time local resident. Small, bowed and thin featured, in addition to his duties as general farm worker, he was also parish clerk, sexton, gravedigger and bellringer at the parish church. He was the only person I have ever known who habitually addressed the boss as 'yes master' and 'no master'.

Charley Butcher and his wife were childless, but they had living with them a lad from London, Don Best, who I got to know quite well during our stay, as he became second cowman during our second year there. He had been evacuated to the Butchers during the early days of the war, he never knew his father, and his mother had disappeared during the blitz. He used to go up to London at week ends from time to time, and he told me that he still hoped to find some trace of her, but so far as I know, he never did. The farm, of something over 300 acres, was divided fairly traditionally between arable and grass, the latter being mainly to the north of the little road running through the village from the main Bury to Thetford road to Livermere and beyond.

Some of the meadows had ponds in them, and the lower ones bordered the little stream that runs from Livermere lake through to Culford and eventually to the River Lark. In addition to milking the cows and looking after the calves for a few days after birth, until they went across the road into the care of the stockman, I also had to keep the fences stockproof round all the meadows. Holes were repaired with barbed wire, the principal means of controlling stock apart from thorn fences, before the days of electric fencers. Bachelor Major Lucas had frequently informed me that barbed wire had been invented by a woman. Whether true or not, it is wicked stuff, and I have seen some horrific injuries caused by stock breaking out through a fence which has had a strand of barbed wire run along it to repair a previous gap.

The Timworth herd in front of the church 1951

When all the milking was done by hand, one man was considered to be necessary to look after fifteen cows, and this number had only increased to twenty or twenty five by the mid-fifties. Thus, when we moved to Timworth I had 23 cows to care for. The following autumn, a large batch of heifers calved for the first time and numbers went up to 52, so Don Best was put in the cowshed as my assistant. Don had a rather flamboyant character which revealed itself through his smoking habits. Payday was Friday, when we all queued up to receive our pay through the kitchen window. Cash in those days! From Saturday to Monday, Don would be puffing away at Churchmans No 1, Players Navy Cut or Wills Gold Flake. Monday lunchtime to Wednesday evening, it would be the turn of Players Weights and Woodbines. Over those five days, he would start by leaving half of each cigarette unsmoked, though the butts tended to get shorter each day. All the butts found their way onto the various windowsills round the cowshed. Thursday and Friday saw Rizla cigrette papers in the ascendency, the butt ends were methodically opened and rolled up anew in an effort to keep going until payday. One week, before harvest, we had been spending our spare time mowing rubbish with scythes round the buildings. Come Friday morning, the last butt end had been used and Don was getting desparate for a smoke.

"Give me a cigarette paper Don, and I'll roll you a herbal cigarette." I said innocently. He gave me one and I set off, apparently for my jacket, but actually into the stackyard where I twirled a convincing cigarette from wilted nettle leaves, tucked the ends in and took it back to Don. Grabbing it, he lit it thankfully with a great big inhalation. The exhalation which followed was, I'm afraid, unprintable.

Electricity had not reached Timworth by that time, so it was still Tilley lamps and oilstoves at home, with the copperhole to stoke for wash days and bath nights. At the farm, things were a bit more modern. A diesel engined generator had been installed in a building near the farmhouse which started automatically when a light was switched on. It had a big flywheel and was all brass and gold lined green paint. It was only powerful enough for lighting however. For milking, we had a Lister petrol engine. Fortunately it was a good natured machine which started readily almost always. Not so the other Lister engine for which I was responsible. This one, in the big barn adjoining the cowshed, could be a swine and caused more than one headache as it drove the pump which brought all the water up that was used in the dairy unit from a deep bore, pumping it up to a very large storage tank in the loft over the feed store. Nor was the engine the only drawback with the water supply. The tank itself was not covered and, on occasions when the taps ran dry, it was necessary to remove the tap and extricate the partially decomposed remains of either sparrow or mouse from the pipe. On one occasion, the blockage occurred in the outlet to the tank itself, and Gerve had to climb into a half full tank of cold water to remove a dead bat which was plugging the exit. This he did with true British grit, yielding nothing to protocol, climbing in complete with wellies, jacket and cap.

The cows themselves were a motley collection of Friesian/Ayshire/Shorthorn crosses, and to provide the next generation, a bad tempered Ayrshire bull was housed in a pen across the road, next to the yard where the cows slept during the winter months. On occasions when a cow had to be served, he was let out into the yard with the cow and had to be separated and driven back into his pen when he had done his job. On one occasion, I was working round the cowshed, and Don was somewhere across the road doing something, probably littering. The next thing I knew, Gerve was escorting Don back to the cowshed covered in muck from head to foot and looking pale and shaken. He also had a trickle of blood running down his face from a slight scratch. It appeared that he had been alone in the yard with the bull, who had taken him unawares and knocked him down. Fortunately for Don, the yard was a foot or two deep in juicy squidge as, when the bull had pounced on the recumbent lad to pummel him with his head, his knees being on either side of him, he had simply pressed him into the slop. Dons cries for help had been heard by Gerve who had driven the bull off with a pitchfork and little harm done. In later years I have not been at all sorry to have all artificial insemination as the keeping of bulls in average farm conditions always leaves a lot to be desired. We have the instance of the attack on father before I left home, to be followed a year or two after the above, at a farm we shall be visiting shortly where three Ayrshire bulls were kept, one of which was notorious for his bad temper. He had had his horns removed because of this, but even so managed to 'get' the Head Cowman one day, despite there always being two on hand when the bulls were being handled, and stock whips being used to keep them at bay. Such bulls, once they have 'smelt blood' have to be sent for slaughter as they will 'kill' again at the least provocation.

The Timworth dairy complex was of a type common at that time. Simply a conversion from hand to- machine milking. The cowshed formed a big 'T', with one end of the top bar joining the barn in which the cows' concentrate was kept. The T itself had a manger running all down one side with a partition between every second stall, with a chain tie for each cow. When converted from hand to machine milking, a vacuum pipe was simply added, running along the front of the standings and above the reach of the cows. There was a brass tap over each pair of standings to which was fixed the rubber pipe which took the vacuum necessary to work the milking unit from the air line to the unit itself. Behind the standing was a gutter which took the liquid slurry out of the shed, and beyond that was a walkway for the cowmen, The vacuum line went through all the milking sheds, across to and through the dairy and into the engine house where it was powered by the Lister engine.

Before milking time, the cows concentrates were taken round the shed in a wheeled bin and divided out by means of a scoop, each cow being allocated so many scoops each according to their yield. When the cows were let in just before milking, they each went to their own standing and would be eating their ration while the cowmen would be going round the shed doing up all the chain ties. Each would then await her turn to be milked. At Timworth, the cows were only in the shed for the duration of milking , after which they would either go back out onto the meadow, or, on winter evenings, would go across the road to sleep in the yard next to the bull where hay would be waiting for them in the racks. On the next farm we shall be visiting, the standings were littered each day and, apart from about two hours each morning while the cowmen were cleaning up and littering when the cows were turned out to eat kale or mangolds on the home meadow, they spent the whole winter indoors. Certainly they did not have a great deal of freedom of movement, but they had ample water and feed and they were always clean, warm and dry, which was more than could be said of some herds that slept in covered yards at that time. The term 'covered yard' can be very misleading. The covered part was such an area that was deemed large enough for all the cows to lay down under cover. The remainder of the yard was usually open to the heavens.

Feeding systems used to be as many and as varied as there were herds. There is probably less variation today as most feeding systems are now automated. Gerve's sytem of concentrate feeding was roughly based on yield per cow, but not on maximum yield. He bought four tons of manufactured cattle cake per month, give or take a little. One fortnight, he would have two tons of Bibby's Dairy Nuts, which were somewhat lens shaped and relatively soft and the cows loved them. The other fortnight, he would have two tons of BOCM dairy nuts which were roughly cubes and as hard as rocks and the cows loathed them to the extent that they would not really clear them up very well. The first two weeks, the yields were up, the second two they were down again. Why did we not have Bibby's all the time? Because the miller who was agent for BOCM, Harry Mudd from Sapiston, was a relative!

We have got the cows in, tied them up and fed them, now they have to be milked. For this we have three portable milking units which comprise a 3½ gallon bucket, wide at the bottom with feet to keep it clear of the ground, narrowing abruptly at the level of the handle to the top which was about eight inches across. To this was fitted a lid with a thick rubber ring round it to form an airtight seal when it was fitted in place by bringing the bucket handle up to form a clamp. The lid had a nozzle through it to which was fitted a rubber pipe to the tap on the vacuum line, and through which the milk was drawn as it was milked from the cows udder. Hooked onto the top of the bucket handle was the milking unit itself which, to the casual observer was simply a welter of rubber pipes culminating in four enlarged rubbers stretched through a metal shell, this being the part that was attached to the cows udder. In the loop of the handle was fitted the pulsator, also driven by the vacuum, which provided the milking source by imitating the squeezing action of the milker's hands, or the calf's mouth for that matter. The vacuum pulsators could easily go wrong. For instance if one of the milking liners split, the milk would escape into the vacuum cavity and be drawn into the pulsator, flooding it with milk. In such a case, the cowman had spares and the know how to soon get going again. Nowadays an electronic pulse unit will control all the pulsators in a milking parlour, perhaps as many as 16 or 20. If one of these pulse units goes wrong they can only normally be put right by an expert from the milking machine agency. They very rarely do, but the time needed to get them repaired is much greater.

After each cow was milked, the milking bucket was emptied into a spare bucket which was standing ready and, as this became full it had to be carried along the various gangways and across the yard to the dairy to be emptied into the receiving tank, no real improvement on the way things had been done at home. The cooling system was also exactly the same, though a somewhat larger model.

After morning milking, the churns had to be rolled outside and humped up onto the nearby churn stand for collection by the churn lorry which, for some inexplicable reason at that time, came all the way from Cambridge. Each churn held 10 gallons and I would not imagine that the Timworth herd would have produced more than a maximum of 150 gallons per day. Memory suggests a maximum nearer 130. Certainly, the cows were not very well bred.

After milking came the cleaning up of the cowshed and washing up of all the milking equipment and dairy utensils. Nowadays, with cows all milked through a parlour of some sort, the former is a quick push through with the squeegee (how ever did we manage before these were invented?) and five minutes with the pressure hose. Then, we had to go all through the various sheds with a wheelbarrow and shovel, and perhaps make two or three journeys out through the barn to an area of foul ground at the back with loads of cowpats. That area was tidied up from time to time by one of the farm men with a horse and tumbril. The floor was washed down with the aid of two buckets, one standing under the tap at the junction of the T while the other was carried to the spot where the cleaner upper had got to, to be hurled onto the floor at an angle adjudged to be doing most good, followed by a vigorous scrubbing with the broom, and a rush back to collect the next bucket of water before it overflowed. Plenty to keep you warm on a day when there were frost patterns on the windows and a sheet of ice outside the door for, of course, there was no such thing as heatng to ease the cowman's lot during the winter. He just had to put on another layer or risk hypothermia.

We did have hot water, especially in the mornings, as one of the first jobs on arrival was to light the fire in the boiler. This was in the engine house near the vacuum pump, with a heap of coke in a leanto nearby with which to stoke it, once you had got it lit. Lighting a coke fire directly from a pile of sticks was not necessarily the best way of improving ones temper first thing in the morning, but once it was burning, it provided us with sufficient hot water to last the day even though the fire was allowed to go out after we had finished the sterilising.

After milking, cold water was sucked through each milking unit individually from a vacuum point in the dairy, followed by hot water containing detergent in the morning, or cold water containing hypochlorite in the evenings. At that time the hot water detergent seems to have been almost universally a white powder called 'Super Kinray'. A final rinse with clean cold water completed the evening wash, but in the morning the cold rinse was followed by steam sterilisation. All the buckets and utensils were put in a large square steam chest, the milking clusters were hung on hooks on the wall with the milk tubes pushed onto nozzles on the steam line. The fire was stoked until the water boiled, indicated when the safety valve started to blow off, a wheel was turned on and for twenty minutes the dairy was filled with hissing steam.

Once a week all the rubbers and milking clusters had to be dismantled, thoroughly scrubbed inside and out and put together again. This occupied a whole morning and, bearing in mind the harshness of 'Super Kinray', left the cowman's hands looking something like uncooked chicken with wrinkled skin. Something of an ordeal if he had to follow this with a job outside in the freezing winter wind. Rubber gloves did not appear routinely in cowsheds until the '70s.

So we jogged on for a couple of years, milking, mucking out, sterilising, helping on the farm in slack periods, keeping the fences mended round the pastures and so on. I had a day off during the week and worked every weekend, a habit which lasted all my working life and, as more people have become mobile, one which I have been thankful for. When we have wanted to go out somewhere birdwatching, there are far fewer people about during the week than there are at week ends, when everybody else is there as well.

Most of the young bloods on the farm were football crazy. On one occasion that I remember, Gerve was away for the day and five or six of them, all the younger ones apart from myself, spent a whole morning playing football of a sort, using a bag tightly tied up to represent a football, and the barn doorway as the goal. Had Gerve found out, I have no doubt that there would have been a sudden dearth of workers on the farm as instant dismissal was the order of the day for misdemeanours of this type. Several of them did actually play football for Ampton, the seat of the 'Squire', Sir Pearce Lacey, who was landlord of both Timworth farms, as well as all the houses in the village, and those in Ampton and Great and Little Livermere. His was a somewhat shadowy presence and to my knowledge, I never saw him.

Football was never my scene, but I did occasionally join the younger members of the farm staff, augmented by other locals, on the meadow next to the railway line where 'cricket' was played during the long summer evenings. Not that I was any good at cricket. My batting was abysmal and my bowling worse. Length and direction were terrible, but I could spin the ball so that it would literally sit up and beg, and in this wise, I could sometimes tempt one of the slashers into striking the ball with tremendous force, only to see it pop straight up in the air and come down gently into a close fielder's hands. One such was John Plume's brother Geoff, who stood over six feet, wore glasses, and handled a cricket bat like a tennis racket. Slashing cross batted at everything, he rarely missed. When he was batting, the fielder at one end of the meadow would scarcely be able to recognise his opposite number at the far end, so far apart would they need to be.

At home, we got on well enough , though the old house was terribly incovenient. The bedroom floors were sloping, the beds tended to be on castors, as was the high cot, so this had to be placed at the lowest point. Susan had a propensity for sitting up and rocking to and fro, quite quickly propelling the cot across the floor and to within reach of things that she ought not, such as bottle of cod liver oil standing on the chest of drawers, which made quite a mess when she managed to get the lid off. When Rachel arrived in September 1950, she spent the first few months in the treasure cot beside our bed, but as our room was very small, she soon had to move in with Susan, heightening the necessity that Susan should stay in one place. There was only an open doorway between the two rooms, so we could hear every move that either of them made. Incidentally, by contrast with Susan's protracted arrival, there were only 3 hours between Rita's first warning and Rachel's first wail. Admittedly she was a bit early and much smaller than her sister. The well in the garden was just a few steps from the back door and for most of the time we were there, the water level was high enough for us to fill the bucket without using the winding gear. The 'little house' was also nearby instead of half a days march away. We had a good sized vegetable garden, with very good soil and no rabbits. The most abundant weed was new to us when we first moved in. This was Lesser Snapdragon of Weasel's Snout, with pink flowers less than a quarter of the size of the garden Antirrhinum to which it is related. Like so many arable weeds, it has become very scarce since the advent of herbicides, never having been a 'dominant' on farm land. It is said to indicate soils of a somewhat acid nature, and the fact that we had hundreds of plants in the garden suggests that there had been a continuing supply of farmyard manure into the soil without a corresponding input of lime to offset the acidity of the muck.

The land on the farm was mixed, ranging from very light Breck soil types such as the 63 acre Bush Breck at the top of the farm on which, each summer, Stone Curlew still nested. I often heard their unmistakable 'Willie Reeve' calls as I was getting the cows up in the early mornings, or late in the evenings from the garden if the air was still and the wind in the right direction. As the crow flew, they were only three miles from Bury St.Edmunds itself. Garden Field on the other hand was clay, this being borne out by a colony of the always rare Blue Pimpernel, now considered to be no more than a subspecies of Scarlet Pimpernel. I have seen the blue form in several places, and always on clay. One plant which we missed while we were in Timworth was the Wild Grape Hyacinth. Not the same as the usual garden species, though sometimes grown, it is believed to be a native in its several sites in Breckland where it is now afforded protection, mostly on undisturbed roadside verges, most, if not all of which are in Suffolk. It so happens that its Timworth site was on the back road which we did not walk many times, and presumably not during the spring period when it would have been in flower.

Woods belonging to the Ampton estate started just round the corner, and today, these are pretty dense and intensively managed. Fifty years ago they had been much cut about and opened up, as an army camp had been built there during the war, and there were the

View from bedroom window, 6, Timworth Green, January 30th 1951

remains of considerable numbers of nissen huts with open grassy areas between. The terrain was ideal for both Tree Pipits and Woodlarks so, even within the confines of an albeit, small village, we could still sit in the garden on a spring day and listen to the lazy, relaxing song of the latter nearby. Nightingales were also much in evidence during the spring and summer and, perhaps surprisingly, Hawfinches were so commonplace that I find a reference in my diary to the effect that they would only receive a mention if the occurrence was of sufficient interest. They may have nested in the Ampton woods. Certainly they did in Culford Park which was only three miles away and toing and froing between the two lots of woods was of daily occurrence. On several occasions we even heard them sing. It is quite a musical though somewhat hesitant song rarely heard by the average birdwatcher, to whom Hawfinches are frequently dumpy 'specks' flying over high in the sky saying 'Tzik'.

The variations in the soils round Timworth can best be appreciated by considering which plants grew where. One of the prime sites at that time was the Green itself. When last seen, this was a wilderness of coarse grass, the dried up pond giving succour to such things as stinging nettles. Then, it was almost heathland, with short turf, and rabbit grazed, at least in part. The only other management it received then came from casual grazing by the cows when they were being taken to one of the short term leys down the Heath Road, but of course, this helped keep the vegetation under control. Among the many heathland plants which grew there Crested Hair Grass was prominent (Koeleria macrantha). Several of the early season Clovers occurred, including Rough, Knotted and Subterranean. All three are of very restricted distribution today, but were rather more findable half a century ago. The same could be said about Fiddle Dock, so named from the shape of its leaves. Hounds Tongue and Wild Mignonette are both plants preferring a heathland habitat, whilst the rarest plant on the Green was actually one preferring a damp situation. There were two good sized clumps of Round-fruited Rush growing on the edge of the pond. I have only seen this in two or three sites ever, and usually fairly near the coast. By contrast, that part of the village known as 'Timworth Heath' which was all under the plough apart from roadside verges, grew several species typical of clay or chalk. These included Wild Basil, Stone Parsley, Fairy Flax and Dropwort. It was whilst walking to and fro across the Green with the cows that I became interested in moths, as several species were met with that I had not noticed before. I was able to obtain copies of the two volumes of South's 'Moths of the Wayside and Woodland' published by Warnes, and this has led to a continuing interest in the group ever since. Among my early discoveries were Latticed Heath and Mother Shipton, both found on the Green. These are species of grass heaths, as were two of the butterflies listed from there. These were Brown Argus and Grizzled Skipper. Of the last named, the Butterflies of Suffolk, published in 1986 states 'The Grizzled Skipper has virtually disappeared from the county, no record having been received of it since 1979.'

Whilst many of our cycle rides inevitably took us back the few miles to Icklingham and West Stow, a favourite long walk with the pram was to Greater Livermere and across the park to Ampton, this being a public right of way, and back down the roads through the woods. In Sir Pearce Lacey's time, this was a delightful walk, through parkland with ancient trees and bordering woods. There was much noisy actitivity in the Heronry on the island in the lake at the appropriate time of year. The huge stick nests were built in Alders and were easily visible from the bridleway. When he died, the estate was sold and the new owner soon proved to have little care for the countryside other than for what he could get out of it. This included felling many of the ancient trees and ploughing and cropping the park. The lake was let for fishing, a formerly unheard of pursuit on an estate where the shooting had been for the gentry and the keepering rigorous. Within a year or two the Heronry was deserted, and since that time, many of the Alders have fallen down. I was fortunate enough to have been granted a birdwatching permit which allowed me to visit one or two of the duck

Livermere Lake in 1952

shooting hides for birdwatching, subject to my observing certain conditions, but now, these became used for fishing and other fishing platforms were erected, and disturbance became too great for the more discerning species. Since then numbers of Canada Geese and Greylags has rocketted. I suspect that these were formerly kept under control by the keeper judiciously taking the eggs. Now the numbers of geese in winter can be in excess of 1000, which is death to any choice plants which could have been growing round the margins, not to mention the masses of 'soup ducks'-Mallard crossed with Khaki Campbells and Aylesburys, with a plumage pattern similar to everything and nothing! The bridleway itself has been much damaged by the passage of farm equipment and would be a pretty impossible route to take a pram over nowadays.

One autumn day, I returned home from a ride round the Brecks to find a message directing me to a cow that had calved on the meadow and needed my urgent attention. Don was supposed to be doing the relief under Gerve's supervision, and I was annoyed. Next morning, I suggested injudiciously to him that if he wanted me to be on call on my day off, he should give me notice to that effect. He replied that if I wanted notice, I could have a week, turned and strode off, the only involuntary move I ever made. Still, I was able to advertise in the East Anglian Daily Times under my own name, and without the use of a box number. I had eleven replies within two days, and the following week end we were off on the next stage of our study of the lowing herd and its environs.

Garboldisham
(Pronounced Garblesham)

We were moved to the Ling on October 23rd 1951 by Will Backlog, the Garboldisham village coalman who also did contracting, particularly for my new employer, Scots farmer John Laurie. We could have gone to his other farm at Ixworth Thorpe which would have been a shorter move, distance wise, but the available cottages on that rented farm were very small. There was the added incentive on the home farm which he owned, that all the cottages had electricity and mains water, something we had not known so far in our married life. At one time, the Ling must have been a series of small farms, as most of the houses,

No.16, The Ling, Garboldisham, 1951

built of the local clay lump, were of small farmhouse proportions and had been subsequently divided into pairs. They were well equipped with outbuildings and stood in enormous gardens by farm cottage standards, which may have lent something to the general atmosphere of the place. Ling Farm was a completely new experience to us. We were not isolated. In fact, we had next door neighbours and two more pairs of cottages beyond them. The other side of us was a single meadow and then the dairy unit, with the rest of the farm buildings and farmhouse nearby. Yet we were isolated in that it was a walk of nearly a mile to the village where was to be found the village shop and the bus route which went from Bury St.Edmunds to Norwich, both places more than twenty miles distant. There was also a market bus on Fridays to Diss, which was seven miles away. Other services were obtainable at East Harling, a large village which served as a small market town, four miles cycle ride distant. The village sign, a lamb on a plinth, testifies to the importance of Harling Lamb Fair in days gone by.

Isolated as it was, the Ling was a community in itself. Thirteen clay lump cottages, mostly in pairs, all of whose inmates worked on the farm, with other workers cycling in each day. At that time, the 500 acre farm supported 33 full time workers, six of which were cowmen. Of these 33, only two were over forty, and John Laurie himself was still under fifty. One of the hardworking and industrious breed of Scots farmers who had come down to carve out a niche for themselves in East Anglia. It was often said of John that when he arrived he had just one set of clothes, and when these became too wet to work in he had to go indoors and dry them out. Whether true or not, I know not. Certainly he had done well in the thirty years he had been in Norfolk, and though a very good and fair employer, he expected every man to pull his weight. Thus, at a time when everything was still done by hand, he grew 120 acres of sugar beet, and every operation from chopping out and singling through to knocking and topping and carting off was done by piecework. The chopping out and singling were done at a relatively slack period in the cowshed, and the cowmen were expected to play their part by doing some in the evenings. All the hedges were kept cut and neatly trimmed. This was also done by piecework, the men being paid so much per chain(22 yards) according to whether there was one year's growth or two. The same applied to ditching and many of the other routine jobs. Though there were tractors on the farm, two full time horsemen and ten working Percherons were still kept, and not only did they open all the furrows and lay all the tops, they also ploughed at least one field close to the main Thetford to Diss road so that all could see the standard of workmanship still practised at Ling Farm.

What did the workers think of all this hard labour? This was the only farm that I ever worked on which had almost if not quite 100% Union membership amongst the permanent staff. By 'permanent staff' I am referring to those who stayed put, rather than the 'itinerants' who, at that time, seemed to include cowmen among their number. Did this mean unrest at work, strikes, lock outs and bad relations generally? No. It meant that the piecework rates were negotiated amicably and properly without any bickering. All the locals were very

Susan with some of the Percherons

much all for one and one for all. Those big gardens would be ploughed at the appropriate time by one or more of them with a pair of borrowed horses and a plough. They would plant them with sugar beet or potatoes, and, as the season advanced, would hoe, horsehoe, chop out, mould up and harvest, each helping the other if the need arose. If the year's crop were sugar beet, all would be harvested and, with a joint permit for the sugar factory, Will Backlog would make up a load from the various plots, each participant being allocated a percentage of the resultant cheque according to his approximate tonnage. I had not heard of the Burston School Strike in those days even though Burston was only about 7 miles away, but there was that sort of community spirit which suggests to me now that, though the strike could not possibly have taken place in such parishes where there was a notable squirearchy, it could have taken place at Garboldisham. Mind you, even there there were one or two who could, and did, rock the boat. For instance, a year or two before, John had sold all the trees from a wood and these had been clear felled and taken to the sawmill. This was at a time when wood clearance for corn growing was still in its heyday. This left all the tops laying about where they had been cut off, and most of them were good, seasoned oak. John told everyone on the farm that they could have all the firewood that they needed and that they could borrow the horse and rulley to cart it home. A rulley was a low , four wheeled wagon with no sides, and with removable high ladders fore and aft. It was especially manouvreable as it had a full lock which meant that, in a difficult corner it could be turned in its own length. I went in with my then next door-neighbour, Arnie, of whom, more later, and, with the aid of his cross cut saw, we carted home two rulley loads each, to be sawn up at our leisure over the winter. Having had a year or two to season and being nearly all oak, it was a jolly good bargain from our point of view. Most of the others on the farm did likewise. Not Halley. He had been in the navy towards the end of the war and for a year or so after. He was the son of a local small farmer and was a grafter to say the least. He frequently regaled us with tales of his naval service while at Lossiemouth with motor torpedo boats, and how they sometimes went out in a small boat on misty nights to some remote point on the loch and passed over a forty gallon drum containing high octane aviation spirit to a shadowy figure in another small boat, for which, he and his companion would receive a fistful of banknotes. On the present occasion, Halley had the horse and rulley on his days off for months on end, and ended the winter with a stack of 35 rulley loads. He then borrowed a tractor and sawbench, cut it all into short logs and sold it. I can imagine John's reaction when he heard about that!

During the next year, I probably learnt more about cows, as well as human relationships than I had in any previous comparable period. For one thing, there were 104 standings in the single covered dairy unit. Four lines of 26 with two gangways, one between each pair of lines, with room for a horse and tumbril to pass down each line for cleaning out and littering. In addition, when numbers were at their peak, we might have up to ten cows in the old milking shed, and these had to be milked by hand. The cows were all pure bred, though not pedigree Ayrshires, and they were much more highly strung than the Friesian 'offshoots' I had so far been familiar with. Of the six cowmen, five would be working at any given time with the sixth on day, or week-end off. The system worked very well, with a different day off each week and a three day week end every sixth week.

At the beginning of each milking, the two most recent acquisitions-myself and Arnie while we were both there-went round with a bucket of warm water and a cloth to wash all the cows udders, at the same time taking the first couple of squirts from each teat into a special cup to check for mastitis. We were followed by two of the senior permanent staff with four bucket milking units each, who did the actual milking.

In turn, we followed behind the milkers with bucket and stool to hand strip each cow of any remaining milk. Most cows would have none, but one or two awkward types would deliberately hold some back until they were stimulated again by the hand stripper, and would maybe let another half gallon or so down. One or two of these were also bad tempered by nature and Arnie would leave these as quickly as maybe, having crouched nearby for a minute or so, pulling nervously at a teat with one hand and holding his bucket in the other proclaiming 'whoa' in a loud voice at frequent intervals. A few minutes later, Fred, the head cowman would come along and notice a bulging quarter into which the cow had let more milk as a result of Arnie's ministrations.

"What's the matter with old 88 Arnie? Looks like she's got some more milk there."

Arnie would trudge back carrying his bucket and stool. He'd put the bucket down and give the offending quarter a poke.

"No Fred! That en't milk. Thass wind, thass what that is."

Halley on the other hand, when he was doing the stripping on Arnie's day off, always claimed that he could sit under any cow in such a way that she could not kick him over. He tried that with old 88 0n one occasion. She was an almost white cow with a wide spread of horns and just a few ruby red markings round her face. Halley sat down on his three legged stool, tucked himself in and started to strip. What followed seemed to be in slow motion. The cow did not kick. She quite deliberately lifted her leg until she felt it tightly under Halley's knees, then she continued steadily lifting so that he went right over backwards and landed flat on his back in the gutter.

The fifth member of the team was the senior person on duty, usually Fred, the head cowman, whose main job was carrying the milk to the dairy. This was a busy job with 8 milking units in operation and all the milking done in about two hours. The set up was just a larger version of what I had been used to, the milk cooler had plugholes for three churns at a time instead of one, and with a set of steps for the milk carrier to climb, enabling him to empty the heavy buckets without the necessity of lifting them up over his head. When one considers that we were producing over 300 gallons at peak times, even with the advantage of having steps it was still a pretty arduous task. When Fred was off, this was done by Putt (real name Percy), who was second cowman, and who would otherwise have been using the milking units on one side. The other regular milkman was Lauder, Fred's brother, real name Harry, who had always had a tendency to sing loudly and out of tune, hence the nickname! Lauder had a motor cycle accident soon after our arrival and was off work for several weeks, during which time his place was taken by Billy Baker, the pigman. Billy only came in at milking times and took the place of one of the strippers, as he had his own charges to see to for the rest of the day so, for five days a week, we were one man short for doing all the things which had to be done between milkings.

Billy was quite a character, and was one of the two over forties. He lived in one of the cottages at the top of Soap Hill, on the way down to the village, with his wife and 12 children. In his spare time, he was much in demand as a shoe repairer, a job he was very good at. He also belonged to the local branch of the county library in the village and claimed to read up to three books per week on top of his other activities. He was probably about fifty, with grey hair and complexion and grey clothes as well, as he always seemed to be well coated with meal dust obtained while mixing the food for his piggy charges.

As the cows were all calved down in the autumn and early winter, there came a time of year when the three bulls were all turned out to grass. We had hardly moved in when, one evening, there came an urgent knocking on the door.

"Can you come and help? The bulls are out."

Three bulls, one noted for being dangerous, one fairly old and crochety and the third young and of untried temper, at large, at night, against a small army of people armed with tilley lanterns and pitch forks or stock whips. Not the most confidence boosting exercise for a new worker, but one that we survived. I only have hazy memories of the occasion, but I think the poor beasts fled back to their paddock when they saw the advancing lights approaching from all quarters, accompanied by a great deal of hallooing.

Another early memory of the Ling cowshed was my first experience of milk fever. The hardy, crossbred Friesians that we had had at home and at Timworth, seemed not to suffer from this, but the Ayrshires did fairly frequently, and subsequent herds have all had their share at some time or another, especially the Guernseys that were my charges for the last nearly 20 years before I retired. Milk Fever is not a disease as such. It is simply a deficiency of calcium in the blood at the time of calving, when there have been heavy demands on the calcium reserves already present. Nor is it possible to take preventive measures as the calving date approaches as the body does not have a storage facility for calcium. We went into the cowshed one morning at five o'clock and there was poor old 52, just recently calved, lying flat out and groaning horribly as if about to die, which she soon would have done had not help arrived. Fred rushed across the yard and knocked John up, who in turn phoned the vet in Diss. Within half an hour, that worthy was injecting 400 ml of Calcium borogluconate into 52's jugular, and half an hour after that, 52 was struggling to her feet. By the time we had finished milking, she was well on her way to being herself again. Milk fever is one of the more satisfying of cow ailments to treat as the recovery rate is usully very rapid. Not always as quickly as the case mentioned, as nowadays, the cowman has his own supply of CBG and his own flutter valve. While some cowmen may inject into the jugular as is the norm with a veterinary surgeon, I always preferred to inject subcutaneously, that is, under the skin. This can be done into the skin of the neck, though Vets are not in favour of this as it may cause a hardening of the skin which is a nuisance when doing the tuberculin test. My favoured position is high up in the loosish skin behind the shoulder blade. This can be done on whichever side the cow is lying, or preferably, half in each side to increase the area of assimilation, and also to reduce the area of the immediate swelling. The needle, which is about four inches long and fairly thick, is inserted through the skin, but parallel with the ribs. The fluttervalve, with its wide end already fitted round the neck of the bottle, has its nozzle end pushed into the back of the needle, and away we go. The liquid should bubble out quite quickly, providing the bottle is held upside down and sufficiently aloft. On the other hand, if the flutter valve has not been washed thoroughly in hot water since its previous use, and is still sticky , or if the needle has only been pushed through the first layer of skin instead of both, the operator may find a very slow passage of calcium into the cow and have a very tired arm into the bargain. Recovery is much less dramatic than with intravenous injection, but this has two advantages. First, the cow can have an injection as soon as she is found, rather than having to wait for the vet to get out to her, and secondly, because the injection disperses much more slowly when under the skin, it also gives the cow extended coverage. This can be important in very bad cases. The exceptional cow will still be down and showing few signs of recovery after four hours, whereupon it is customary to give a second bottle. The first dose would be Calcium borogluconate 40% but the second bottle would more usually be CMPD 20%, Calcium, Phosphorus, Magnesium and Dextrose. Such cows needing a second bottle may get up fairly quickly thereafter, or they may stay stubbornly down. Further bottles may be prescribed by the vet, though occasionally, she may be better in herself, but too weak in her back to make it to the upright position. In which case the vet may suggest a course of TLC (tender loving care) which, with plenty of patience, could see her up in a few days or even a week. If not up in that time, it may be decided to bring in the hip clamp and lift the cow bodily to her feet. This is done by placing the adjustable clamp over each of the projecting hip bones, screwing it up tight and gently raising cow and clamp using a tractor foreloader or forklift for the purpose. It may be necessary to repeat this performance for two or three days, until the animal has finally had sufficient exercise for her seized up muscles to be able to get her up and down unaided. On the other hand, I have found that a cow which gets to the stage of having to be lifted has got less than a 25% chance of recovery anyhow. Not just from injured back or stiff muscles, but also from liver damage caused by the severity of her attack. If she does die, or has to be humanely destroyed, the knacker is the only answer.

After milking came the clearing up. At the Ling,the two 'juniors' mucked out with the horse and cart, usually Arnie and I, while he was there. After mucking out, we had to take the horse and rulley along to the strawstacks beside the road leading from the Ling towards the main Diss road. We had to load up with as much straw as we could pack on, especially during the winter, when we had to have enough for all the cow's standings, but also for the calves and young stock in the old buildings. The cows on the Ling, were very comfortable, as they had ample bedding provided, far more than is used in some cubicle houses nowadays, though of course, the straw appears less nowadays because it has been compressed into bales instead of being carried in just as it was threshed. They were always much cleaner in their shed than they would have been in a covered yard, and on a very cold morning, it was a real pleasure to get the milking units in throught the sliding doors and close them again, thus keeping the cows overnight warmth in and the temperature at a bearable level in which to work. Those who have milked in a parlour without the benefit of a wall heater will know just how cold they can be. The cows went out onto the home meadow for a couple of hours while we were mucking out and bedding down, even in the middle of winter. While we were doing this, Putt would be going round with trolleys of food and putting the required amount of scoops of this and that ready for each cow when she came back in. The food consisted of freshed pressed sugar beet pulp during the season, which was followed after the beet factory was closed by dried pulp which had to be soaked with water for a few hours before use. This was topped with concentrates. At Ling Farm, cattle cake was not used. Instead, we had a 'coarse mix' which was very palatable, and consisted of various cereals and oilseed cakes blended together with molasses to prevent it being too dusty. The cows milked extremely well on it too.

Part of the Ayrshire herd, Ling Farm, Garboldisham, December 1951

The meadow gateway became desparately muddy during the winter, with only the one narrow entrance for upwards of 400 cloven hooves to splodge through whilst going in and out. One day, Lauder had been sent to get them in. The cows all came in well enough, though rather dirty, but Lauder was a long time arriving. Presently, we became aware of somone swearing at the top of his voice, so we all rushed out and gathered round to watch. Lauder had been trying to scurry across the worst part as quickly as possible to avoid sinking in, had misjudged the terrain, and one of his boots had been sucked off by the clinging mud. Due to his speed at the time, the boot could be seen with the top sticking out, a full five yards behind the spot where Lauder had come to rest, just about up to his knees in mud. It did not help his temper particularly to have an admiring audience as he struggled round and got back to his lost boot, to make his way 'ashore' carrying it in one hand.

In a way Putt, though only second cowman, seemed to be one of the most important men on the farm at that time. He was about 35 and lived at home with his parents and unmarried sister. His father was a retired blacksmith and Putt was skilled in many fields. On one occasion, he was taken out of the cowshed for a couple of days to show some of the other men on the farm how to lay a pipe drain, a much more skilled operation than might be imagined today, when it is all done by machine.

During a slack period in summer, he had a fortnight out to make a considerable number of clay lump bricks in the traditional way in order to repair some of the cottages and outbuildings on the farm. I had often wondered about the very deep ditches round the meadows, now I learnt how they came to be so deep. When clay was needed for making clay lump, instead of having a pit which would take up valuable space, they simply dug the ditch bottom a little deeper. A large tilt, or stack cloth was laid out on the grass next to the ditch, and a layer of clay was thrown onto it from the ditch bottom and spread out evenly. Long wheat straw was added, and with the help of a little water to make the clay malleable, the straw was trodden in until it was fully incorporated, using Bonny, the cowshed horse to speed the process. This system would not seem to have changed much since the Israelites left Egypt! When of the correct consistency, Putt filled as many wooden moulds as the mix would fill and left them to set. Within a day or so, they were turned out and left to harden in the sun, but needed to be covered up if rain threatened. Clay lump is extremely durable providing that it has an adequate covering of plaster. Plastered inside and out, it is notable for its warmth in winter and coolness in summer, as I can testify, having lived within clay lump walls for our year on the Ling, and again between 1960 and 67 as we shall see later.

Putt was always an easy going chap, so it was therefore remarkable that it fell to him to 'settle' Arnie, and this he did in a very short space of time. Arnie was a truculent little character. He was just forty, and he had been milking cows, according to his story, since he had left school at fourteen. Twenty six years, and the longest that he had stayed in one job was two, with quite a number of 'workstops', one could hardly call them otherwise, of three months or less. He was married to a lady fourteen years his senior, and one who had probably had a strict chapel upbringing, if the illustrated texts which adorned their walls were anything to go by. They had a daughter who at that time was eleven, so her mother must have been forty three when she was born, and a very bad time she had had at the time, so she told Rita on one occasion. They also had a female lodger, a short, somewhat overweight lady of about forty, of very plain appearance, who was reputed to have already had two children, though never married, and who was currently working as a domestic at a boy's private school at Garboldisham Manor. Arnie himself was about five foot three, and with a very inflated view of his own importance in the system. He would strut up and down the cowshed with his chest puffed out like a bantam cock and fingering his watchchain, when he was not deliberating and commenting on this and that like a sage of old. Say anything amiss and he would go up like a rocket. Though he wore thick, horn rimmed spectacles, he would fight anyone at the drop of a hat , and in a tough, knockabout environment like the Ling was at that time, he got plenty of egging on. Such as the time when the lads persuaded John Laurie's son Jimmy, then about fourteen or fifteen, to bring his boxing gloves down to the cowshed one afternoon while father was out, and challenge Arnie to a match. To say that the result was hilarious is something of an understatement. Jimmy had been taught to box properly at school, while Arnie knew nothing of boxing science and cared less. He was also a good six inches shorter than his opponent. I do not think that Jimmy threw a single punch. He just warded Arnie's attacks off as the latter came leaping in, both arms flailing wildly like a demented Jack-in-a-Box with the crowd gathered round, cheering him on. I cannot remember, but assume that the match ended when Arnie became exhausted.

More sinister for the peace of the cowshed was the vendetta that erupted between Arnie and Halley. The latter took my place on days off in the littering and mucking out process, so I never knew exactly what sparked things off, but can imagine. Bickering led to open hostility until one day, when Halley had to be forcibly removed from on top of Arnie, where he had him pinned across the shafts of the rulley behind the horse. Arnie then did the unthinkable and went round to the house to tell John how he had been ill treated. 'Creep arsing' to the boss for whatever cause was anathema to all the men on a farm like the Ling where, as I have already said, it was 'all for one and one for all.' Unfortunately for Arnie, John Laurie was the one farmer who I ever worked for who would not listen to talebearers for whatever cause. The following afternoon when I was back at work and had heard part of the story and, like all of us, was wondering what would happen next, the sliding door at the end of cowshed opened about two feet and John's head came round.

"Next time there's any fighting in the cowshed," his voice boomed out, "both parties get the sack." The door closed leaving two parties to mull over his words.

A week or two later, and the farm foreman, Peter Baker, usually known as 'Tairter', was having a few days holiday and John and his wife were out for the day, so Lennie Bailey, the head horseman, was in charge. Lennie was also Putt's great crony. After breakfast, Putt came to me in a state of great agitation. I never knew what sparked it off, but he was breathing fire and brimstone.

"Alec, will you do the feeding and washing up this morning? I'll go with Arnie and muck out and get the straw. I'll settle that little ole bugger for good!"

Now Putt was big and broad, and strong as an ox. The mucking out was done in record time, Arnie having to scramble to keep up with his side of the gangway, and off they went to get the rulley load of straw. The straw was loose in the stack and it was necessary to cut a slice across with a hay knife before throwing it down onto the rulley a forkful at a time. One of the tractor drivers stopped at the cowshed on his way into the yard to ask what was going on, as he had seen Putt rolling great heaps of straw down onto the rulley, with Arnie shouting and waving his arms about as he tried to keep on top of the onslaught. Soon after, Putt came striding into the yard leading Bonny and a load of straw that looked anything but secure. Arnie should have stayed on top of the load to pitch it down but, at the first opportunity he came sliding down the rope breathing fire. He rushed round the rulley, fists at the ready, to have a go at Putt, who simply stepped aside, grabbed him by his collar and the seat of his pants and headed for the horse pond. At that point, Lennie Bailey hurried across and persuaded Putt, after some altercation, not to drop him in as the pond was deep, and, in all probability, Arnie could not swim.

Arnie saw the writing on the wall this time. Putt was one of the most respected of the permanent staff. Within a few days, he had found another job and given his notice in. The day before they moved, the news leaked out that the lodger was expecting her third child. Arnie had been taking advantage of his wifes' long walk to the village to take the child to school, to indulge himself at breakfast time.

We were sitting having breakfast next morning, when we were startled by a tremendous crash outside the end kitchen window. This looked onto the neighbours back yard and was always kept curtained. A second crash followed, so we hurried across to the window and peeped round the curtain to see what was going on. We knew they were moving that day, but had not expected them to throw the brass bedsteads out of the upstairs windows. No wonder they were all tied up with string!

So ended Arnie's three months on the Ling, and also our first three months, as we had both started at more or less the same time. Things settle down after that, the new neighbours being of a quieter disposition.

As the days lengthened-Arnie must have left at the beginning of

February - we started to look around us a bit more. We had a lot of snow in March in that year, 1952, but nevertheless enjoyed an early spring. By now, Susan was three and a bit and Rachel eighteen months so, to aid mobility, we bought a seat for Rita's bike for Rachel to ride in, and I made a somewhat larger wooden seat and fixed it to my carrier as Susan was a very big child for her age. At the same time, it turned out that I had the better bargain as Sue would sit relatively still and contentedly watch the world go by. Rachel, on the other hand, was always on the move. If she spotted something that caught her interest, she would throw herself suddenly sideways to get a closer look at it, and Rita had to be constantly on her guard to anticipate such movements. Nevertheless, during the summer of 1952, we estimated that we cycled more than 1250 miles round the lanes and byways of south Norfolk and north Suffolk with the children on the back, and I filled three and a half note books with the accounts of our trips.

That was the year that Garboldisham Heath was ploughed up. Today, it is a series of arable fields running from south of the Thetford road right through to the East Harling road, behind the village. The last time we went that way, the part of the former heath to the south of the Thetford road which runs down to Garboldisham Fen and the Little Ouse beyond, was planted with second rate grass and even the old marl pit was much eroded, by comparison with the area it formerly covered.

With a whole day off each week and a three day week end every sixth week, we had a fair bit of scope for our cycling trips, one of which was as early as April 11th, a very warm, sunny day for the time of year.

Our first stop was on Garboldisham Heath, a typical stretch of Breckland grass heath, especially in the vicinity of the old marl pit, but also with interspersed areas of acid heather heath. As we freewheeled down the hill, our first Tree Pipit of the year was singing from a tiny elder near the road. Nearby, a male Wheatear was indulging in frequent, bouncing leaps into the air in song flight, looping sharply back each time to drop down near the point that he had just left which was a telegraph pole. Even when he did not land back on top of the pole, he went back to it for his next upward flight. The gentle song of a Woodlark came to us from a point further from the road, this being suddenly blotted out by the twittering of 26 Meadow Pipits which leapt up from the nearby heather and circled around us. These were undoubtedly a late winter flock passing through. The Garboldisham parish boundary lies just at the bottom of the hill, the boundary bank being one of the 'Devil's Dykes' of Breckland, a prehistoric earthwork of uncertain age and purpose. This one runs more or less intact from the Little Ouse marshes to Harling Fiveways and, before it became overgrown with bracken, was one of the few sites for the very rare Spiked Speedwell, one of the Breckland rarities. Earlier this century, it was frequent over a fairly big area of the heath as the late W.G.Clarke, author of the Breckland classic 'In Breckland Wilds', visited the heath with his friend W.H.Burrell of Thetford and wrote that they found many thousands of plants of the Spiked Speedwell. Whether this was just not found by us, or whether it had, by that time, been overcome by invading bracken, I cannot say. Certainly we only explored that part of the heath that was relatively near the road because of the smallness of the children. Beyond the Devil's Dyke on the south side of the road, and in Gasthorpe parish, there were one or two meadows, from the first of which we now heard the unmistakeable 'Willie Reeve' of a Stone Curlew. We quickly spotted a pair strolling about on the grass though the calling bird seemed further away. The pair eventually got up and flew across the road, to land among the heather not far from where we were standing.

A Sand Martin circled us singing his indeterminate little song, and I twice thought I heard a bird call that I did not associate with heathland. I soon realised that I was correct when two gorgeous male Yellow Wagtails came dipping across the heath and landed just a few yards away, immediately starting to potter about among the heather looking for insects which they seemed to find plentiful. We strolled across to a large tumulus on the north side of the road and in so doing, discovered there to be at least three or four pairs of Tree Pipits and two of Wheatears on that small part of the heath alone.

On June 8th we went specifically to make a list of the plants to be found growing there, and it was as well that we did. Though the area adjacent to the Devil's Dyke seems to have been known fairly well by botanists, I would question whether the whole of the heath had been studied closely. Starting down by the Fen, we found such plants as Common Meadow Rue, Marsh Cinquefoil, Marsh Stitchwort and Marsh Speedwell, before moving up to the disused marl pit. This had a little water in the bottom and a strange assortment including water loving plants like Common Spike Rush, mixed with a variety of chalk loving plants including Fairy Flax, and these living cheek by jowl with plants of acid habitats such as Hoary Cinquefoil and Trailing St.John's Wort. Crossing the road to the north side, we eventually arrived at a shallow depression from which chalk must have been taken at some stage, and which then presented us with a bewildering array of chalk loving plants including Fern Grass, Purple Milkvetch, Rock Rose, Salad Burnet and Squinancywort. Last, but by no means least, we discovered the pit area to be starred throughout with the bright yellow flowers of Spring Cinquefoil. I took a specimen home to identify and fortunately pressed it as it is the only record for this species in Norfolk, though Simpson's Flora of Suffolk records that Hind in his 1889 Flora of Suffolk had it from Icklingham, Cavenham, Risby and West Stow, with an 1889 record from the last named the final record for that county. Spring Cinquefoil is more typically a plant of the northern limestone, and we have seen it in quantity around Dovedale in Derbyshire. Within a few months, all these plants had disappeared before the bulldozer and the plough, and with them, the Grizzled Skipper butterflies, or rather, the next generation caterpillars of those that we recorded that day, and the habitat of all those heathland birds. Even the Sand Martins were nesting in the old marlpit, though this by that time was far from suitable. Two pairs had their nest holes in the narrow band of topsoil above the ancient face of poor quality chalk on that day in June.

The horsepond at the farm deserves a mention other than for threatening to drop recalcitrant cowmen in! It covered probably a quarter of an acre and was of very pure water, as was the nature of ponds before the days of intensive fertilisers and pesticides. It contained several species of Pondweed including Curled and one or more of the grass leaved species which I was not very good at at that time. It is the only farm pond that I remember which had nesting Little Grebe on it. On one occasion, I was fortunate enough to be able to watch them at close quarters while I stood in the tumbril in the middle of the pond. At this distance of time, I have no idea as to why I should have been standing in the tumbril in the middle of the pond, but there I was. Suddenly, the male Little Grebe came swimming out from the bank towards me dragging a big mass of vegetation with him, to a point about 25 yards from where I stood. The female was already there, apparently floating 'at ease'.

As the male arrived, she snatched his load from him and dumped it unceremoniously just in front of her. This was followed by such a flurry of activity that could scarcely be thought possible. Both birds started diving and hauling large pieces of pondweed to the surface, to be dumped higgledy piggledy on top of the first piece. Occasionally one or other of the pair would take a longer dive to bring up an especially choice piece of plant debris from the very bottom of the pond. In this way, they probably 'visited' the nest ten times per minute each, and by the end of five minutes had accumulated a sizeable mound of rotting vegetation, which is all that the nest of a Little Grebe consists. The eggs are laid on top of it, and when the bird leaves the nest to feed, she just pulls some of the weed over them to deceive the prying eyes of Crows and Magpies. I did see some of the eggs after they had been laid, and could make out the pattern on them, but within a very few days they had become stained and discoloured by the nest material.

On our way to Scoulton Mere

Going back a few weeks in time to May 1st, we set out on our first really long distance cycle ride with the children. On that occasion, we went north, and out of Breckland altogether. Our first, and main objective was Scoulton Mere which lies beside the road from Watton to Norwich and about seven miles east of the first named. Watton was famed in the Garboldisham area for its weekly market. If you had not been to Watton on a Wednesday, you had not really lived! Many were the tales of the bargains available on the pavements of the High Street. Having said that, the market still thrives to this day, and is still held all down the High Street even though this is the main through route from east to west. The town has one advantage over any of its neighbours in that the High Street is straight enough and wide enough to accomodate the market without seriously impeding the passage of traffic even at the present time. Having said that, if I have to go 'Watton way' on a Wednesday I usually find a route round, rather than going right through the middle.

The morning was taken up with our journey from the 'border country' near Garboldisham, through countryside varying from woodlands near Quidenham and Eccles, semi-heathland round Hargham (Harfum), and open arable round the Ellinghams, until we at last reached the Scoulton area, where we immediately found a beautiful shrub that we had not seen before. This proved to be Bird Cherry, considered in many books to be of northern distribution, but quite common in parts of central Norfolk and the Broads district.

As we drew nearer to Scoulton, the numbers of Black-headed Gulls increased, especially round an Ellingham duck farm where they could find easy feeding. Reaching the main road near Scoulton school, we cycled along it past the Hemsworth Arms to the point where the Mere was marked on the map. Like so many village pubs, the Hemsworth Arms has long been closed. Most have become private houses though in this instance, it became a 'pine shoppe' for a number of years until the business outgrew the premises and moved to the then recently closed school, and the former pub then became a private house. Reaching the Mere, we were very disappointed to discover that the island marked on the map as 'the Heath' was completely covered with Alder, with just a few Mallard visible on the water. We turned round and went back to the last road junction, where we turned into the minor road signposted 'Woodrising' and after a short distance, again came within sight of the Mere. At the north end of the island, the trees disappeared and were replaced by lush marsh grass. On the water were between 150 and 200 Black-headed Gulls whilst around them, and around the low part of the island, dozens more wheeled and squabbled for position. On the island itself, we could see scores of heads peeping through the long grass, telling of birds sitting on their nests. They appeared to be so closely packed that there was scarce room for another to land between those already present. A few were also sitting on nests on the bank of the Mere in a small overspill gathering. The whole area was protected by deep ditches and barbed wire fences. At that time, Scoulton gullery was the largest in East Anglia and the tenth largest in the country.(Handbook of British Birds, First Edition, H.F.Witherby 1940) Over the preceding centuries and well into the present-some

Scoulton Mere, May 1st 1952

sources say until the beginning of the 1939-1945 war-the gulls eggs were collected on a vast scale each spring, with 20-30,000 a year being sent to London, where they were sold as 'Plovers' eggs'. Real Plovers eggs were those of the Lapwing, also known as the Peewit. In Norfolk, the Black-headed Gull is also known as the 'Peewit Gull', so the extension of the name to cover the very similarly shaped and marked eggs would seem a logical step if you have a regular source of eggs! However, White's Directory of 1845 has really got hold of the wrong end of the stick when it claims that-"Scoulton Mere, a fine sheet of water two miles in circumference has in its centre a small, swampy island. (It is actually nearly half the total area, the Mere being the outer boundary) and is frequented from March to July by many thousands of Peewits(Lapwings)which are useful for clearing the land of insects, and also servicable to the poor who gather their eggs for sale." The bracketted reference to Lapwings reveals that White had the wool pulled thoroughly over his eyes. Neither were the eggs collected by the poor, but by the estate, who made a substantial sum each year from their sale on the London markets.

Following our visit in 1952, a chapter of accidents befell the gullery and nine years later when we moved to a village only three miles away, we visited the Mere to see how the gulls were getting on, to discover that the many centuries old gullery had disappeared, probably for ever.

Another frequently visited site was South Lopham Fen, as this was only four miles from the Ling. South Lopham lies on the north bank of the source of the River Waveney, whilst Redgrave lies on the Suffolk side, with important Fen habitat on both sides of what, at that point is no more than a ditch which spends much of its time dry for the first two or three hundred yards. The road that crosses the upper end of the Fen is the watershed, and water falling on the east side runs down the drainage gulleys into the ditch and eventually reaches the sea via Breydon Water and Great Yarmouth. Rain falling on the western slope of the road runs into the ditch which is the source of the Little Ouse which, after a much longer journey round south west Norfolk arrives at Denver Sluice in the Fens where it is released into the tidal water through the sluice and eventually reaches the Wash north of Kings Lynn more than 70 miles from Yarmouth. In 1952, the whole Fen on both sides of the Waveney was open, with much Saw Sedge which in places was almost impenetrable. Saw Sedge was traditionally much used in Norfolk by the reed thatchers as it is very tough and durable. It is used for the final decorative ridge capping, the part of the thatch which is subject to most wear by the weather. To maintain a healthy stock of reed and sedge, both of which were harvested at South Lopham, all the dead material had to be eradicated periodically, and this was done by burning late in the winter, before any new growth had taken place. On April 14th, I cycled down in the evening and recorded in my diary that a pall of smoke that had been visible from the Ling for the previous two hours, was coming from a fire that was burning right across the Great Fen, the wind being in the east. By that time it had moved into Middle Fen but appeared to have broken up into a series of smaller fires, flames from which were shooting to a height of twenty feet or more. One arm had burnt itself back against the wind to the head of Great Fen. I then spent an hour or so listening to two Nightingales, stalking and eventually seeing one of two reeling Grasshopper Warblers, and noting that a Woodlark was singing along the edge of the Redgrave side at the Little Fen end. This was completely open apart from isolated aspen and birch near the entrance from the road, where was a small area of open heathland with Heather and Needle Whin. As I was leaving the fire had flared up again, presenting rather a terrifying spectacle in the dusk as it advanced in a long crackling line towards Little Fen, but a sudden shower which drove me to shelter, put the blaze out in just a few minutes.

Talking to Fred, the head cowman, next day, I learnt that this had been a traditional firing of the Fen to burn off the dead herbage. So far as I am aware, that was the last time the Lopham Fens complex were deliberately burnt off. There has been much discussion about the merits and demerits of controlled burning some holding the view that it does a great deal of damage. Obviously it does some damage, but is it long lasting? The Lopham Fens had apparently been deliberately fired for years, probably generations, and they provided a very rich habitat which supported a goodly number of rare plants and insects. It was years later that the discovery was made of the Great Raft Spider (Dolomedes plantarius) in the Fen, at that time, its only known site in Britain, so it must have survived innumerable burnings as it is hardly likely that it could have arrived from anywhere else. Lousewort, Butterwort, Sundews, Grass of Parnassus and several species of Orchids flourished there in 1952 and indeed, though we never saw it there, the very rare Fen Orchid was recorded there until 1957. By that time the firing had ceased and the Fen fell into neglect. Aspen, Birch, Sallow and other shrubs started to take over and within a few years, the whole Fen was well on the way to becoming a vast swampy scrub. The newly formed Suffolk Trust for Nature Conservation, now the Suffolk Wildlife Trust, stepped in and managed to acquire the whole of the Redgrave/ South Lopham fen complex as one of their earlier reserves. They had a tremendous uphill struggle halting, and eventually beating back the invading scrub, though the damage had already been done as the herbage had become much coarser due in part to drying out. This situation was greatly exacerbated by the presence nearby of a new borehole sunk by the East Anglian Water Company, and in addition, a number of other local bores for spray irrigation. However, they were able to get the Fen designated a RAMSAR site under the European convention and the Water Company eventually lost their case and had to resite the borehole that was causing the worst of the drying out. Considerable damage was done, especially to the flora by the drying out which unfortunately coincided with three years of drought, and experts decided that this had resulted in the top layer of peat having become sterile. The chosen option thereafter was to strip 6 inches of peat from 80 acres of the fen which was most impoverished, raising the water table by means of sluices and reseeding the area with litter cut from a similar, though much less rich fen a few miles away at Market Weston. Much of the area is now fenced and, after experimenting with grazing cattle, management is now carried out by grazing with a free roaming and breeding herd of Tarpan ponies from Poland which are reputed to be experts at scrub elimination once the trees have been removed. All very well in theory but not so good when one considers all the factors involved. Fen management in the past may have involved tethered animals such as ponies or goats grazing limited areas of drier ground round the margin of the fen. Traditional management of fen areas was for a variety of causes. Some fens had rights to gather sphagnum, or limited rights for cutting peat as fuel. Lopham and Redgrave were a thatching resource, but grazing free range by cattle or horses, and even with sheep except perhaps the odd two or three, would not have taken place except on the drier areas. For one thing, as wet as fens formerly were, there would have been a continual risk of animals getting stuck, so, even if the dried areas were being grazed, they would have been in charge of a boy whose job would have been to keep them out of the wetter areas. In addition and as I think we may well see in the fairly near future, enrichment of the fen environment with cattle, or even horse dung will result in the impoverishment of the flora which will be exacerbated by the cutting up of the fen surface by the hooves of horses or cattle breaking through and fragmenting the fen mat upon which the rare plants and their dependent fauna are reliant.

On the other hand, grazing by sheep on such a reserve as that at East Wretham heath is fully justified by tradition. The heath includes the two well known stretches of water, Ringmere and Langmere when they are not dry, as they have been with increasing freqeuncy in recent years, due no doubt at least in part, to water extraction for spray irrigation in a land where it is now fashionable to grow potatoes, sugar beet and carrots on blowing sand. The track that goes from east to west between the two meres was formerly known as the Harling Drove, and it was along this that flocks of sheep used to be taken from inland Norfolk to their summer grazing on the Fens beyond Hockwold. Though the Drove disappears today where it meets the Peddars Way, it formerly carried on until it reached at least as far as East Harling.

The Peddar's Way is the Roman Road that started in Essex, though there is some dispute as to where in Essex, and runs right through Suffolk, passing through my home village of Hitcham and crossing the Little Ouse at Knettishall Heath. Today, the Way can still be followed along most of its original course from Castle Acre, north of Swaffham, right to the coast at Holme next Sea where it may have been an embarkation point for Roman traffic seeking to bypass the treacherous fenland basin en route to their bases in Lincolnshire and further north. The Meres were a common watering place , not just for passing flocks, but also for neighbouring parishes. Long fingers of land stretch up to Ringmere from the parishes of Croxton, Wretham, Bridgham, Kilverstone, and Roudham , and a similar finger from Brettenham stops a little short of the mere, but the people of Brettenham were allowed

Langmere, April 23rd 1952

right of way for their flocks to go to Ringmere to drink.

In spite of this picture of much rural coming and going, the meres were formerly a haven of peace. Well, more or less. Certainly they were at the time W.G.Clarke was writing 'In Breckland Wilds'. But even Clarke admits to the area having been a metropolis of a sort when he

Langmere, January 1987

wrote "Here, for over half a mile the 'Drove' is a parochial boundary between Croxton and Wretham and soon passes south of Langmere and half a mile north of Ringmere,(it is actually somehat less than this). The abundant flint implements by the meres indicate permanent settlement by prehistoric man, and his tools are widely scattered on the adjacent Brecks and on the whole course of the track itself." And again, "Langmere is the most beloved of wildfowl of all the heath meres, and this despite the fact that its margins are devoid of any vegetation which could shelter bird or nest."

This is not true today as the heath badly scrubbed up following myxomatosis and the temporary demise of the rabbit which allowed birch and gorse to proliferate where none had been in Clarke's time. Writing this early in 1999, I can say that the meres are just starting to fill up after the longest period that we can remember them being dry. That this has been partly due to several drought years there can be no doubt, but the real cause for this lengthy absence of water can only be spray irrigation. Unlike other stretches of water, the meres do not respond directly to rainfall. They often stay stubbornly dry through a period of prolonged rainfall, and will suddenly start to fill up three or four months later when everywhere is dry following the passage of the previous rainfall to the appropriate level in the chalk aquifer. Above Langmere is a long mound which, when the mere is full, becomes an island. More usually the mere is divided into one large pool and two smaller ones. Crowning the ridge though now surrounded by the scrub mentioned, are a handful, and now rapidly diminishing as they die and are blown down, of ancient pines. When we first visited the meres, they stood proudly and healthily up above the mere, seven or eight in number and, to quote Clarke again 'said never to have been planted by the hand of man.'

On July 11th I wrote, "As we arrived at Ringmere, the whinneying of adults and incessant feeding calls of innumerable juvenile Little Grebe was the predominant sound. Several families were seen, as were those of Moorhen and Coot. A family of Lesser Redpolls circled overhead as we left our bikes and strolled across toward Langmere. Here, a continual rumble and clouds of dust denoted the passing of tanks from the nearby training area. This seemed not to worry the birds in the least. Many Meadow Pipits and Yellow Wagtails and a few Pied Wagtails were sweeping back and forth over the water's surface picking from the swarms of insects, which would have been caddis flies and similar species. Higher flying insects were being snapped up by myriads of Sand Martins and lesser numbers of Swallows and House

Martins. The mere is now low, with a ridge dividing the main pool and the one whose margin we were following. Though again, the air was filled with Little Grebe calls, here, these were outnumbered by the calls of Lapwing, several young birds of the species being visible-round the water's edge. As we approached the ridge,12 Teal got up from the nearer pool. A bird was spotted walking cautiously and quite quickly away, and eventually flew. What a setting in which to see a Stone Curlew at close quarters." There followed an account of what we saw whilst peeping over the top of the ridge down into the main pool. Apart from duck, this included 6-8 Redshank and 7 Ringed Plover. The latter is a coastal wader which at that time still nested quite widely on the bare parts of Breckland heaths. We had been lucky enough the year before to spot a bird sitting on 3 eggs only a short distance from the road between Thetford and Barnham. On a later visit, we actually found a Stone Curlew's nest and eggs on the island at Langmere, surely one of the last times that it could have nested there. Following Myxomatosis, things started to go radically wrong. The general area of the meres became quickly grown up with Birch, whilst the island at Langmere was soon invaded by Gorse and Broom. Thetford became a place of London overspill with a rapidly increasing population looking for somewhere to go on a Sunday afternoon. The Heath had been partly purchased by the

Stone Curlew's nest, Langmere island, August 1st 1952

Norfolk Naturalists Trust and partly gifted to them in 1938, though whether the rights of common access for watering purposes had been extinguished, I do not know. Sufficient to say that a crisis was reached during the '70s when Sunday invaders from Thetford were said to have numbered more than 1000, complete with inflatable dinghys and all the paraphernalia of modern living, and they did not take their litter home. The NNT(now Norfolk Wildlife Trust) reacted by building a house and installing an ex-Naval officer as warden. He was followed by a second warden who was even more insensitive, and at the same time, the whole area was shut up except to permit holders who had to pay for the privelige. In more recent years, there has been a relaxation of attitudes, not just at Wretham, but widely throughout the country now that the County Wildlife Trusts and other similar organisations have come to realise that, with wildlife such a high profile subject on television, more and more people want to enjoy the countryside with as little restriction as possible, and more and more people now belong to the county trusts and organisations such as the RSPB. At Wretham, management is carried out using the Trust's 'flying flock' of sheep which are rotated round the reserves, and this is reasonably satisfactory, though whether they are resposible for the diminuition of the area of heather which used to be extensive to the east of Langmere, or whether this has died off following an attack by the Heather beetle, a creature which operates in the same way as the elm bark betle, responsible for carrying Dutch Elm disease throughout the countryside, I do not know.

Among other places of interest close to Garboldisham were the forest rides at West Harling, not far from Harling Fiveways and only a few miles from home. In 1952, many of the blocks of pine were still quite small and the track not too churned up by forest traffic in spite of it being a 'public road' through to Bridgham, with an old bridge over the River Thet about half a mile beyond the most interesting part of the track.

At some time, following current forest thinking at that time, all the top soil was scraped off to the depth of two or three inches, thus, in many places removing all the vegetation down to the bare chalk, the very shallow subsoil at that point. Whether this was thought to be a ready made fire break, or whether it was an attempt to level the track to facilitate forestry operations I do not know. Botanically, it was a disaster as it removed a nice piece of chalk grassland with such regulars as Purple Milkvetch, Rock Rose and Dropwort and also a small colony of Spanish Catchfly, one of the Breckland rarities that has always been

West Harling Old Belt, now Bridgham Lane picnic place. September 6th 1952

scarcer in Norfolk than the Suffolk Breck. In addition, a small colony of Moonwort was wiped out. This strange little fern has only been recorded in a few places in East Anglia, being more usual on limestone in other parts of the country. The only other places that I have seen it, many years ago now in both places, was on Lakenheath Warren and near Two Mile Bottom on the Thetford to Kings Lynn road, in an area now long under mature pines. A good many years later, a second similar soil scraping exercise, this time levelling an area to store tree trunks as a compartment was clear felled, wiped out the largest colony then known of the Field Southernwood, sometimes rather clumsily called Breckland Mugwort. The first time we saw this Red Data Book species was on August 3rd 1952. We cycled to Thetford and onto the Bury road to Barnham Cross Common to search for it where it was mentioned as occurring in 'In Breckland Wilds'. I noted in my diary that the Common was very chalky, and in spite of much Gorse and Broom, we found masses of another Breck rarity, Sickle Medick with, here and there, its frequent hybrid with Lucerne which can be any colour between pale yellow and blackish purple by way of green. Here too were numbers of chalk downland plants, with Wild Thyme, Harebells, Stemless Thistle, Burnet Saxifrage, Kidney Vetch, Quake Grass, Eyebrights, Knotted Pearlwort and much more. Eventually, we made our way back to the road and turned toward Thetford. Just before we reached the north end of the common, I went up a track to the west, while Rita strolled down the opposite side toward the ruins of the old nunnery with the children. When I was nearly to the top, I came upon a series of shallow pits and, as I looked down into the first of these, I spotted three large plants with shrubby stems and finely dissected leaves, and I knew that I had found what I was looking for. Field Southernwood still grew where Clarke had known it, for a few years after 1952, but under the Thetford overspill scheme, houses were built right to the edge of the Common. Some years later, after a period when it had not been within cycling distance, I went again to the Common to look for the Southernwood. I discovered that the whole common was still undeniably a public open space, but that north end where the plant had grown had been levelled and was unnaturally green for Breckland, with football pitches and playing fields of other denominations. Incidentally, Whinchats also nested there in 1952, but this species appears to be yet another in terminal decline as a breeding bird. The Suffolk Brecks had about 15 singing males in 1997, whilst Norfolk has not had a single confirmed breeding record anywhere in the county since 1992.

Most of the Common is still relatively undisturbed. There is rather a lot of Gorse and Broom, but open areas are grazed spasmodically by tethered ponies and somebody, presumably the local council, do some cutting of herbage in the autumn which allows most of the chalk grassland plants to flourish, so one would hope that most of the butterflies that used to be there are still to be found.

Looking back through my nature diaries for 1952, I realise that there is more than enough material to fill a book on the wildlife of Breckland. Between January 1st and October 27th, when we left the area, I had filled 470 pages of narrow feint ruled exercise books. How has Breckland changed since then? Almost everywhere we went, from Foulden Common in the far north west to East Harling Common in the east, and all the Suffolk places mentioned earlier which we visited again, there are mentions of Stone Curlew and singing Woodlark. One evening when we were coming home after a very long day, right down to Icklingham and Risby, we crossed Garboldisham Heath as darkness was falling. Rachel was probably asleep on the back of Rita's bike, but I mentioned that Susan was thrilled to hear the song of a Nightjar for the first time. At least two were churring close to the road with a third audible in the distance, where in all probablity, no Nightjar will ever be heard again.

I remember that evening particularly as, when we arrived home on the Ling at past 10 o'clock, both girls were suddenly very wide awake and spent some time rushing round to the furthest recesses of the garden shouting at the tops of their voices, and the neighbours were in bed and no doubt trying to sleep!

Before we move away from thc Brecks, I must mention one more place that we visited quite often. This was East Harling Common, a place I went in search of again just a few years ago. The lane leading to it, which is almost a mile long, was just penetrable on foot. It is a public footpath cum bridleway, but one that did not appear to be well used, even though there are houses for some distance down from its beginning on the outskirts of the village. Gaining the Common at last, I soon discovered that the former ponds were all bone dry, though admittedly this was during a drought period. Apart from a few knolls capped with coarse vegetation, with a few sprigs of spindly heather poking through the bushes, it was obvious that somebody at some time had had a go at 'improving' it by ploughing and reseeding, with the result that 'floral' areas were now a mass of cocksfoot tussocks, ryegrass and other undesirables, and the stream now had nothing to recommend it whatsoever.

I first visited the common on the evening of July 9th 1952. After a very bumpy ride down the track, I left my bike and climbed over a stile onto what immediately struck me as being a very interesting looking area. I soon discovered that it was being grazed by a few sheep and cattle and a horse, which must have been the optimum management necessary to keep it as it was then. In 1952 it consisted of a number of ridges of chalky boulder clay with frequent pools and ditches, some of which were either wholly or partly dried up, that year also being a dry summer. I discovered large quantities of Eyebright flowering on the ridges, with strips of Heather between. This is a typical phenomenon in Breckland where, following the last glaciation, the ice sheets scoured grooves in the top of the underlying chalk. The grooves filled with soil over the centuries, but the soil only barely covered the chalk ridges. This has resulted in quite large areas in places where there are strips of chalk grassland, the soil over the chalk ridges, interspersed with strips of heather, this growing over the grooves where the chalk is too deep to affect the acid loving heather plants. Round some of the pools, the Blunt-flowered Rush was so luxuriant as to be four feet high. The flora was very complex with the whole range of plants from calcicoles to calcifuges mixed together in wildest profusion. Thus there was Squinancywort and Stemless Thistle growing mixed with Gorse and Heather and accompanied by Common Marsh Orchids. Strawberry Clover was growing near one of the pools. Sometimes found on boulder clay, it is more usual on clay banks and pastures near the sea, and virtually never in Breckland. I found Red Rattle, a relative of Lousewort which is quite common in the valley fens of East Anglia. Large quantities of Common Cotton Grass fringed many of the pools, its long white plumes blowing to and fro in the breeze. The characteristic pinky umbels of Water Dropwort were everywhere to be seen, and occasional Snipe, Mallard and Teal were disturbed from the swampy areas. At that season, all must have been nesting nearby. A Reed Bunting was singing its uninspiring song and a pair of Lapwing were in residence. The biggest bird surprise, however, came when I strolled across to a dried looking area and disturbed a pair of Stone Curlew which flew off calling. At this distance of time I can only assume that they had flown in to feed and drink from Overa Heath, just across the nearby railway line, as that had not been ploughed up at that time. Reaching the stream, the infant Thet which joins the Little Ouse at Thetford, I walked the length of the common along the bank. There was an alder carr at each end. The water was beautifully clear and, as there were around the pools, there were swarms of Damselflies, *Coenagrion puella* and *Ischnura elegans,* as well as a number of Common Darter dragonflies. I followed a swimming Water Vole for some distance until he darted into a hidden hole under the overhang of the bank. Reaching the second carr, I discovered a flowering clump of Greater Water Parsnip. This stately umbellifer is on the 'Scarce Plants Register' for the British Isles, though it is not infrequent in the Norfolk Broads, one of its main national strongholds.

Another wet ditch was found to be filled with Marsh-Arrow Grass, Brookweed, an unlikely looking plant to be a member of the Primrose family and Lesser Water Plantain, along with a Sedge then new to me which proved to be Bottle Sedge, so called because of the shape of its ripe fruit. The surface of many of the pools was liberally dotted with bright yellow flowers, clusters of which stood up above the water on longish stalks. These were all inaccessible from the bank, and all the more mysterious until I came to a pool which was partly dry, and filled with the plant. Hauling some out for examination, I discovered that the underwater parts were not rooted in the bottom, but were free floating, the roots ending in bladder like processes, one end of which was open and fringed with hairs. Common Bladderwort obtains its nutrients from Water Fleas and other small aquatic creatures which swim into the open mouths of the bladders, the hairs close behind them and they are trapped. The plant is a close relative of the Butterworts which live in boggy places such as South Lopham Fen and trap insects on their sticky leaves, the edges of which curl over the trapped insect until such time as digestion has taken place. This family may not be as well known as the Sundews with their sticky tentacles carrying out the same function in a more spectacular manner.

Bladderwort is still findable in some of the less polluted pools and dykes in the Norfolk Broads. In 1998, I discovered a 50 yard stretch of dyke with an estimated 5000 blooms standing above the water-and I had left my camera at home! I remember once seeing this plant from the train window when we travelled along the Thetford and Swaffham line in the vicinity of Stow Bedon Common or nearby Cranberry Rough. That particular branch line was one of the first to be dismantled in the post war purge of uneconomic routes, and though parts of the course of the line have long since disappeared, the section mentioned now forms part of a county council walk known as the Great Eastern Pingo Trail. What, you may ask, is a pingo? It is actually an Eskimo name for a phenomenon which occurs in areas of permafrost, in which a lens of subterranean ice forms, as during the Ice Age, in an area of peaty soil. Perhaps thousands of years later, as the frost goes out with the amelioration of the climate, the lens will melt, the peaty soil over the upper side of the lens will collapse into the hollow formerly occupied by the ice, and a water filled hollow in the subsoil will result. It seems that this was most likely to have taken place where the subsoil was chalk. There are a number of areas of pingoes round the Breckland margins, several of which are nature reserves or SSSIs, such as Thompson Common, just beside the Pingo trail, East Walton Common, Pentney Common, Foulden Common, Marham Fen, and I would not be surprised if the list ought also to include East Harling Common.

If we seem to have spent more time touring south Norfolk and north Suffolk over the past year, and very little with the lowing herd, this was very much not the case. A lot of the corners had now been rubbed off which had existed due to my having been brought up in a relatively sheltered existence with very little communication with the outside world. At that time, the owner occupier and his family, however small his holding, was in as far removed a situation from the farm workers as he was from the gentlemen squirearchy who owned the big estates. Even those who owned their own farms felt notably superior to those who were only estate tenants as they had greater security, though this might not have been noticeable from the outside. For four years now I had been rubbing shoulders with the farm workers on equal terms, and I had soon learnt that it did not do to put on airs and graces. My year in the Ling cowshed had taught me as much about human relationships as it had about cows, and that was a considerable amount. As the time off was generous, I had also learnt more in that year about the wildlife of the countryside around me than in probably any year before or since. We had also learnt a great deal more about pulling together as a family which has stood us in good stead to last us a lifetime. But it was time to move on. I was not unhappy at the Ling and neither was Rita, but at 25 and with a growing family it was time to look ahead and think about the future. So I made a mistake. I applied for and got a job as Head Cowman of two looking after a herd of 35 pedigree Friesians at Debenham, in mid-East Suffolk, halfway between Stowmarket and Framlingham. The lads all stood outside the cowshed and waved us off as we went past with our belongings loaded onto a lorry belonging to Lists of Debenham, and I kept in touch with Putt for many years until his somewhat early death during the 1980s, even though it was only a letter at Christmas. Fred, the Head Cowman, died of lung cancer before he reached 50. Not too surprising as he smoked four ounces of Churchman's Counter Shag a week rolled into cigarettes, but then, Putt smoked an ounce of Darkie Shag each day! Not too many years later, after John Laurie died of a heart attack, the herd was sold, the farm became 'modernised' to the extent that it became all arable and was run by the former foreman and three or four of his sons, and was farmed from one of John's sons farms several miles away. All the other workers were made redundant. There is nothing worse for a farm worker past fifty than redundancy as there is little chance of his ever finding work again. Lauder was lucky as he was still young enough to get a job as tractor driver on a farm at Honington, near Ixworth in Suffolk. Halley was alright of course. His wife was a Scot from Glasgow, and the last I heard of him, he had moved to that city and set up as a builder and decorator.

Today,the Ling seems a desolate and empty place to what it was in 1952. We went that way a year or two back and all the cottages have now been turned into 'desirable residences', though the one we lived in looked less well cared for than some of the others. It may have been something to do with the loss of the brick and flint wall round the front garden, the felling of the lime trees by the gate, and the visual impact of a sky television dish on the front wall. The farm buildings are much larger and more extensive than we rememered them, but all were closed and silent. No people, and not an animal in sight.

Debenham

The mistake became evident as soon as we arrived in Debenham. On this occasion, there was no house available on the farm. My new boss, Commander C.J.Carr, had got an option on a council house on a new estate on the outskirts of the village. I had been to see him a couple of months previously and had made all the arrangements for our move. We had had to wait for the completion of the house, but everything would be in order by the end of October, we had been assured. It was not. The house was finished, but it was just as the builders had left it, so our furniture had to be packed in the rooms as much out of the way as possible, and everywhere had to be scrubbed through to remove muddy footmarks, cement dust and everything else that builders and decorators leave behind them. The surroundings were typical of building sites, though nowadays, topsoil is usually spread over after levelling. On this occasion, the site had not even been levelled and, after several days of rain,was a sea of yellow clay and puddles. We soon learnt that, though we got on well enough with people in general, provided that they were not on top of us all the time, we did not much care for a semi suburban environment where it was necessary to go for a fairly long walk to see anything other than two or three seedy sparrows. I was alright as the farm was a mile and a half bike ride away, through the village and up a long steep hill to Crows Hall. Rita had to be there all the time.

Crows Hall, Debenham, August 1953

Crows Hall itself was a marvellous place. Romantic, I suppose some would call it. An Elizabethan moated house, well kept, with a Tudor bridge over the moat and gardens on two sides, the other two being water. There was about an acre of land within the moat, which was beautifully clear and absolutely packed with Rudd. No fishing was allowed, but on one occasion when the Commander and his wife were away, the foreman turned a blind eye to a couple of lads he knew from the village. They came up with their rods for an hour or two, catching several that were fifteen inches or more in length.

The huge brick barn-I did roughly measure it once by stepping it-was contemporay with the house, and in the days when local knights kept their own small armies, had been used as a barracks. It was about 272 feet long and on two floors, and on the upstairs floor were still to be found, several old swords said to date from the time of Cromwell. One end of the building had only a single floor, and here the Commander had installed an early grain drier with six or eight storage silos, connected by a series of underfloor traces and above floor elevators for transferring grain to where ever it was needed. There were two main destinations other than for 'internal distribution'. One was the bagging off point as, by late 1952 the bulk haulage of corn was still very much in its infancy so, when a load of corn was required for sale, it all had to be run into sacks for loading onto the lorry.

The other destination was to the end of the first floor which was open toward the silos. Here, I had to bag off barley, oats and beans for loading into the hopper of the hammermill to be ground up as concentrates for the cows and young stock. As it went through the mill, it was blown into an attached storage and mixer hopper to which minerals and protein supplements could be added and thoroughly incorporated when grinding was finished. The mixture was then bagged off ready for carrying to the milking parlour and shooting into the three feed hoppers, from which it was measured by using a long handled scoop as each cow came in. Yes! We were modern at Crows Hall. We had a yard and six standing abreast parlour. Six cows came in to start with. Three were fed and washed and the foremilk taken, the milking clusters being put on immediately. The second three were prepared for milking, by which time, in theory, the first three should be finished, and all that remained to be done was to was to take the clusters off, release the milk from the glass jar and put the clusters on the second three. Let out numbers 1-3 while 4-6 were milking and let in 7-9 and wash them etc. No more buckets. No more lugging the milk to the dairy. As the cows were milked, the milk went into an eye level pyrex glass jar which stood at the back of the feed hopper on a scale. Of course, this meant that we had to record the milk at every milking instead of once a week as hitherto. This had been done using a spring balance hung at strategic points round the cowshed as we had worked our way round. Once the milk had been recorded, a master tap was turned and it disappeared out of the jar into a large bore milkline which went through the wall into the dairy. Here, it flowed into a brass cylinder with a heavy weight at the back end. As the cylinder filled with milk, it would gradually counterbalance the weight and drop until it cut off the vacuum supply which held it sealed. A flap at the front end then dropped open and the milk poured out into the receiving tank, the weight immediately lifting the cylinder back into vacuum. The 'dairy' end of the process was still exactly as it always had been, as was the washing up and sterilising. Now, I did have a bit of extra knowledge that helped. Putt had shown me how to keep the boiler fire alight overnight by filling it up with coke last thing and covering it over with ashes. This meant instant hot water at all times with very little extra expenditure on fuel, a nice warm boilerhouse first thing in the morning, and toasted cheese sandwiches for breakfast every morning as it was too far to go home.

Something that I did learn about cows at Debenham was that the term 'Pedigree' had little meaning if it did not also allude to careful breeding and selection. The 'Sherrydan' herd of pedigree Friesians were no more and no less than cull cows from a number of no doubt reputable herds at that time. Some of them looked well enough, but only two could manage six gallons a day when freshly calved, and one of those had been disposed of by her former owners because she was such a slow milker that she took fifteen minutes to milk in the morning and ten at night. Nowadays, with factory farming methods, any cow taking more than four minutes to milk is likely to be on her way quite quickly. I also discovered, during the winter months, how much filthier yarded cows could be than those kept in a cowshed. The Timworth yards had not been too bad in this respect, but those at Crows Hall had a much bigger area open to the sky and a very rough, poor, stony bottom which made them almost impossible to clean out properly.

A long avenue of oaks led to Crows Hall from the point where the designated public bridleway branched off and continued down to the Debenham to Framlingham road. At the end of the avenue stood a pair of cottages, one of which was home to Commander Carr's farm foreman, Jack (would you believe it?) Driver. The other was occupied by Harry Exton, driver of the crawler tractor deemed necessary to do all the ploughing on the farm's heavy boulder clay acres. Harry's daughter was married to one of four brothers who worked on the farm during the time I was there. Bob was also a tractor driver, a younger brother, Dennis, had been reluctant milkman before we arrived, showed me the ropes when I started, and was also to do the relief milking. The youngest brother, always known as 'Pip' was to be my assistant. Pip was a quiet lad with a severe disability. He did not like getting up in the morning. We did not start work at Crows Hall until six, and, for a few weeks, Pip arrived at five past-ten past-gradually this became half past. The Commander warned him to mend his ways and this he promised to do but,'when he was a layin' in bed, his back and legs became so rigid he had a job to get them down onto the floor.' This state of affairs continued until he was arriving at ten past seven, by which time, I was sometimes finished milking when he got there. The Commander reasoned with him, more in sorrow than anger, and once again, he promised to do better. Next day, it was seven twenty five and he got the sack. He was replaced by his eldest brother, Bernard, who had formerly been head cowman, but had been attracted by an offer from another farmer almost in the village. Bernard had soon found that he had less affinity with this dour Scot than he had thought, and was glad to get back with the Commander even on a lower level. We got on well together and things went much more smoothly after his arrival. He and his wife lived with her father who was landlord of Debenham Cherry Tree'.

Though more mechanised than most farms at that time, the payroll still included around a dozen men including the two cowmen on about 300 acres, an unthinkable total today. Among these was George, who later became relief milker.

The Commander stuck his head somewhat further out of the door, listening.

"Ah! I hear old Wells is at it again!" I grinned and agreed with him.

That was the thing with George. He could probably have been an opera singer if he had had his voice trained. He had a good singing voice which he exercised at maximum decibels whenever the opportunity presented itself, and especially if he was doing relief milking. He also had one rather less favourable idiosyncrasy, a rather inflated opinion as to his own importance which sometimes led him astray. After nearly four years as cowman, I had discovered that, around the cowshed, the head cowman's word was law and, providing he kept within the bounds of good management, even the boss would not dream of rocking the boat by interfering in any of his decisions. After my day off one week, I returned to work to find that George had been poking around to see what stocks we had of this and that. Not his province at all. I had a regular weekly spot with the Commander during which I put in a list of things that we should want during the coming week. George wrote himself out a longish list of things he thought we might need at some time in the nebulous future and presented it to the Commander. I was furious,as there was nothing on the list that we needed immediately. Bernard, I suspect, thought that I might need a bit of egging on. Bearing in mind that he had previously been Head Cowman, it may even have been that he had a score to settle into the bargain. He certainly filled me in with as much detail as possible, and suggested that I ought to 'go after George'. The latter was ten years my senior, some bit taller than I was and very well built.

That afternoon, we had to go down the Rookyard Meadow to collect the cows from down near the Framlingham road. Haycarting was taking place just through the hedge beside the bottom driftway, and George was pitching onto the trailer actually beside the hedge.

"You want to go after him!" Bernard said for the umpteenth time. I started to berate him at the top of my voice for poking his nose where he had no business to. I'd let him know who was Head Cowman, and so on, as I scrambled down into the deep ditch and up the other side to push my way through a tiny gap into the hayfield. I looked around for my victim. No one in sight. I rushed round to the other side of the trailer to where the other pitcher stood, grinning all over his face. There was George at my feet with his hands over his face. "Mercy!" whimpered a small voice. Now I knew why I had been egged on. It's called 'over reacting' nowadays. George made sure to keep on the right side of the Head Cowman after that.

The farm also supported two Billies. Both were middle aged. One was single and as irascible a character as one would wish to meet. The second was married easy going, and an inveterate rogue. He was employed, partly as a general farm worker and partly as farm maintainance man. He was a skilled carpenter and had formerly had his own business. Watching him over a period of weeks, putting up new guttering round the barn, I could understand why he had not succeeded on his own. He was fond of quoting the old saw 'measure twice and cut once' which can be time consuming when you are working twenty feet from the ground, and each measurement means climbing down and moving the ladder! The twice also seemed to run to three or four measures per cut. He was a craftsman nevertheless and still turned out hand made thatching ladders in his spare time that were the equal of any that I have seen.

Any of the men who wanted milk for their families, came and collected it from the dairy as they were going home in the evenings. At that time of day, Jack,the foreman also collected his milk and stayed a few minutes leaning over the half door for a chat. One winter evening, Billy the carpenter collected his milk and went off, leaving Jack leaning and chatting. Two or three minutes later, there came a loud bang at no great distance, down the avenue of oaks. Jack glanced over his shoulder, then looked back at me.

"Old sod! That's another pheasant gone. You can't catch him, not without searching him. He always carries a folding .410 in the poachers pocket of his donkey jacket." On this occasion, a pheasant as well.

The two Billies were ever at each others' throats. They and a third member of the farm staff, Tom Pallant, were all members of the village fire brigade. One evening, fire broke out on a farm about a mile up the road towards Kenton, and the brigade was quickly on the scene. Billy the carpenter gave the appearance of directing operations, while the other Billy directed the hose. "Over here!" cries the carpenter, to be nearly knocked off his feet by a jet of water. Not the way to douse a farm fire!

Though the birdlife an the Debenham area at that time, still seems to have been plentiful by today's standards, it was somewhat restricted in quality. Near our council house, one of the few noteworthy occurrences was of a flock of over 200 Skylarks feeding on the field opposite. This had remained uncultivated as it was earmarked for a further extension of the housing estate. At Crows Hall, it was a pleasure to find Kingfishers to be of frequent occurrence. I wonder if they are still there? They seem to have declined disastrously over the past two or three decades along with a lot of other bird species. Herons too, were frequent visitors to the fish filled waters of the moat, but for the most part the birds seen were those to be found in any agricultural area with reasonable hedges. To find more, one had to go to the coast or estuaries, the nearest of which, the Orwell, was about fifteen miles distant, but on the other side of Ipswich. The Deben, Alde and Blyth were further afield, though even the last named was only about twenty five miles. During the winter, cycling trips with the children were not really on, and by the time things started to warm up again the following summer, Rita was pregnant again and having rather a bad time. This meant that we did very little travelling around in the

Debenham area as we had at Garboldisham. I did make one or two solo trips that winter, as far as the Blyth, but fortunately not near the weekend of January 31st/February lst-and I would remind you that the year was 1953. On the former date, I recorded in my diary that it had been a day of severe gales, which had continued into the night. On February 1st, the radio reported the strongest winds ever recorded in Britain at that time, reaching 125 m.p.h. in Orkney. The north westerly gale was accompanied by very high tides, breaching the east coast sea defences in more than 500 places, and over 300 people lost their lives, with things even worse in Holland. That day was a Sunday and, during the afternoon, John Denny called to see us. He had cycled over to visit Buss Creek just outside Southwold, before going over the Bailey Bridge to birdwatch on the marshes south of Walberswick, but found that he could not get through as the sea had broken through the shingle beach and joined up with the Creek, making Southwold into an island. This major disaster had a profound effect on the coastal strip around East Anglia for many years to come. The salt water flooding of low lying ground resulted in the appearance of many salt loving plants where they had never been seen before , while some scarce plants with a more or less coastal distribution had a brief flowering in great profusion in places where coarse grasses had previously squeezed them out. During the following autumn, I cycled down to Manningtree, and on marshes near the railway crossing of the Stour valley, discovered large areas relatively devoid of grass, but thick with the flowering plants of Slender Hare's Ear. This small member of the Umbellifer family is an annual which only appears on bare ground in a semi coastal habitat, so one assumes that viable seed must remain in the soil for very many years awaiting an opportunity to germinate when conditions are favourable. Some other areas which have been inundated by sea water on a number of occasions over the years, have a curiously mixed flora with plants of inland marshes growing side by side with those which prefer a measure of salt in the soil. For instance, on the Hickling National Nature Reserve which includes Hickling Broad itself, the largest stretch of open water in the Broads National Park, there is a considerable quantity of Sea Rush, normally found on the higher levels of salt marshes and among dunes where there is sufficient moisture. Water from Hickling Broad flows down the River Thurne to Breydon Water and Yarmouth by way of the River Bure, a journey of fourteen miles. Hickling Broad is, however, only three miles from the fragile line of sand dunes which keeps the sea at bay at that point. I cannot find any mention of the sea breaking through into the Broads area other than at Sea Palling in

1953, and that is further round the coast. However, a major inundation which remained over that part of Broadland for several months during the early part of 1938 was one of a series of such floods recorded back to 1287 when history records over 100 people drowned in Hickling village alone.

The Debenham area was typical of boulder clay country which has been well farmed, even allowing for a number of bankruptcies during the 1930s when the value of land fell to £5 per acre. There was little waste land, or even any marshy meadows or woods of distinction. Aspall wood, in the next parish, was within walking distance and had a footpath running along the bottom which we followed on a few occasions. It was a nice wood with a certain amount of coppicing still taking place, with Primroses, Wood Anemones, Dog Violets and clay indicator species such as Hairy Woodrush. On a roadside verge between Debenham and Aspall, at a point where the road does a Z bend where it had formerly ducked through a bridge on the old Mid-Suffolk Light Railway, we found a colony of 72 Bee Orchids flowering that summer. As is the nature of the plant, though we stopped there on a number of occasions subsequently, we never found it there again.

Crows Hall had some old meadows flanking the infant River Deben. These were fairly unremarkable and probably improved, their only claim to fame being some quantity of the rare Wood Clubrush growing along the banks of the stream. The best natural history site locally was undoubtedly the moat surrounding Crows Hall farmhouse. I wonder what it is like today? It was a wonderful haven in those days, with clear, unpolluted, fish filled water supporting the larvae of Caddis flies, Damsel and Dragonflies, the perfect insects of which emerged in succession during the late spring and summer, and a wide diversity of pond weeds and waterlilies, some of which may have been introduced. Plants of old walls festooned the sides of the moat near the house, and across the bridge, and especially the lovely introduced Pink, *Dianthus plumarius*. These all looked as if they had been there since the house was built. Looking through my notes, I see that Shepherd's Needle was listed from the stackyard without special mention, so it was still a common plant on boulder clay at that time. Now listed as a rare and declining weed of arable land, along with a number of others, it was one that I grew for a number of years in the 1980s as part of a rare arable weed project initiated by Suffolk Wildlife Trust. I still have this and a number of others from the project maintaining themselves in one of the vegetable gardens including Corn Bedstraw. Very like Goosegrass or Cleavers to look at, the tiny white axillary flowers are borne in threes, and as the seeds approach maturity, the pedicels turn downwards giving the plant the appearance of bearing a series of miniature pawnbrokers' signs. The only time I have ever seen this in the wild was in our Debenham 'garden' and all over the uncultivated field across the road.

The old windmill at Debenham, August 1953

Living on a new housing estate which, even after a whole winter, was little more than a sea of mud, and with neighbours close on either side of us and all around us, some of whom we got on well with and others not, confirmed our preference for the solitude of open countryside. Our attempts to grow a few flowers and vegetables on a garden wholly devoid of topsoil could only be described as pathetic. Rita became pregnant as spring became summer and was very ill for several weeks. As I was nearly two miles away at work most of the time, she and Rachel went to Hitcham, to my parents for several weeks, whilst Rita's Aunts at Westhorpe had Susan for a long holiday. Rachel later went to stay for a while with her other granny at Little Clacton. At about this time, my Grandmother Bull died, and at the end of the summer, Grandma Somers, now 82, decided to give up her bungalow before another winter and move in with my parents. With her bungalow were some buildings and an eight acre field of shallow clay.

Here, I could amuse myself in my spare time and help father on the family farm for the rest of the time. For the next seven years, my only contact with the lowing herd was a house cow which I had to milk by hand. Toward the end of the time we were there, during which I still had secret hopes of restarting a milking herd on the farm, I did have three calves which I bought on the open market. They were alright for a couple of days, but one of them, a little Jersey, could not take the sudden change from rich milk to the then current milk substitute which contained a variety of milk solids plus various vegetable additives like crushed linseed, and on the third day, I found her dead. This about summed up my spare time efforts. We had hens in deep litter, which were all the rage at the time, the desired hours of extra daylight in winter being provided by a couple of Tilley lanterns. Soon after we had the young birds- they were bought at eight weeks old- something frightened them and they all piled into one end of the shed, about a dozen dying of suffocation. Later, when fully grown, several developed the habit of cannibalism and had to have their beaks clipped to prevent them pecking each other to death. Rats became a plague until I 'discovered' Warfarin rat poison, and I was able to surmount that problem. I had a cross bred in pig gilt from father who produced a litter of 13 and brought up 11. These did quite well, but were ready to go to market when pig keeping was in a trough, something that happened less often than it does at the present time. Having produced the one litter, the gilt went down with kidney trouble, wasted away and had to go for slaughter.

On the field, I grew barley and wheat, and an acre of sugar beet. How I hated that acre each year! The field had more than its share of couch or Spear Grass as we called it, which made hoeing the beet nothing but sheer drudgery. In hot dry summers, it was like hoeing a concrete road. If it was wet, half the field stuck on the hoe blade and the rest on my boots, and by the time I had them lifted, knocked, topped and carted off in the autumn, everything had become cold and muddy and horrible. I have never had the least desire to be an arable farmer, and what desire I might have had was effectively stifled by that acre of sugar beet.

Meanwhile indoors, Rita played her part manfully looking after the needs of the family, which had been joined in January 1954 by Sally and in February 1959 by Christine. Four girls, the eldest to be 12 in late 1960 and the youngest two in early 1961. All needing food, clothes, schooling, as children always have done since schools were invented! Even at that time, it was more than a family farm of under one hundred

Hitcham. Rita harrowing the field, April 1954

acres could support, for father and mother also had sister Gillian to school, she being 16 years my junior.

In the late summer of 1960, I started looking round, went for what seems now like innumerable interviews before I found the right place, or so I felt, and in the October of that year, on a pouring wet Saturday, one of the wettest days I can actually remember, we packed all our goods into a lorry and trailer and set out the forty odd miles to Cranworth, in Norfolk, where I was to come face to face with the lowing herd with a vengeance.

Cranworth

Innumerable interviews? Was it so difficult to get a cowman's job in 1960? No, it was not. There were still plenty of jobs available to suitable applicants. I did not get the one nearest to home admittedly, but that was because the post in question was that of Head Herdsman to Lord Blakenham, then Sir John Hare, as he wanted somebody with experience of Channel Island cattle to look after his herd of pedigree Guernseys. I do not think that we could have gone there anyhow as, though a very nice tied cottage with all mod cons was available, the three bedrooms would only have comfortably housed ourselves and three children. That was the reason why I went for so many interviews. Most of the cottages were too small. Rita came with me as far as was possible to make sure that I did not take a job where other considerations were totally unsuitable. Very wise of her, remembering Debenham! We were beginning to wonder if anything was going to come of our searching, as I had advertised, as well as replying to other people's ads. My latest effort, in the East Anglian Daily Times, brought an out of county reply from E.R.Eglington Ltd of Booter's Hall, Cranworth. It did not sound too far away, as the post town was Thetford. One of those curious quirks of the post office that the

Church farmhouse, Cranworth

Cranworth and a good number of other villages in that area,mail has always gone by way of Thetford, as a consequence of which the village comes within the Ipswich post code area, yet it is only 7 miles from Dereham and 16 from Norwich, whilst Thetford is 25 miles away.

This time, I had to go on my own as it was going to be an all day journey going by rail from Stowmarket to Norwich, changing to the Dereham and Fakenham line to get off at Thuxton station. I was met by Mr Eglington in his landrover, and discovered him to be an affable gentleman who had some difficulty walking. I later learnt that he suffered from arthritis. He showed me round the herd and the milking parlour, and more importantly the house. Church farmhouse at that time was divided into two, with his son in law, who was foreman on the farm in one end, the other being for the Head Cowman. It was an eighteenth century farmhouse built of plastered clay lump, and the cowman's half had four bedrooms, though one was quite small with a leanto roof on both sides, being in an extension. That did not matter. There was room for everybody. Having been shown round, I was taken to the farmhouse and regaled with roast beef and yorkshire pud plus trimmings, followed by plum pie and custard.

I was nearly taken unawares at the start of the meal when Mr Eglington said Grace before eating. However, I did not commit any gaffes and, as I was being taken back to the station, I was told that the job was mine if I would like to take it. I took it.

I was somewhat heart in mouth as we splashed through puddles and heavy rain all the way up through Suffolk and South Norfolk. What a day to take ones wife and family to start life in a completely new area, in a house that they had never seen! Had I done it again? When we arrived, it was to find that Erica Clayton, our neighbour, had got a good fire going in the kitchen range, and later, came in and helped Rita make the beds up after I had helped the lorry driver and his mate carry all the furniture in and place it roughly where we wanted it. Things got a bit wet between the lorry and the house, but nothing suffered permanent damage.

This was an important move for the older children. Sue had already started at Hadleigh secondary modern school, so was automatically transferred to Watton secondary modern in whose catchment area Cranworth school lay. She was not of academic bent, but Rachel, now 10, would be taking her 11 plus in a little over a year and looked to have a promising future. What was Cranworth school like? It was almost opposite Church Farm and was ruled over by Mrs Zanella, of whom, everybody spoke very highly. We need not have worried. Mrs

Zanella was everything people said that she was and all the younger girls prospered in turn when they went to Cranworth school.

What about the cows? 75 pedigree and grading up Friesians to be cared for by yours truly with the help of a second cowman. While they had been short handed, a relief milking service cowman had been looking after the herd for about six weeks. What about the second cowman? Oh! He only started work last week! The day I started work, he had his first weekly half day off. We worked a six and a half day week for the first two years I was there, and I was shown the ropes by Ken, the relief milking service man. He later came back to work as Head Cowman on Mr Eglington's other farm at Bawdeswell, where a second, similar sized herd was kept, but after a year or two, moved on to become a company rep.

To help the new cowmen get to know their charges, their names had all been printed on saleyard type labels and stuck on their rumps with saleyard type glue, so all would be well and we would soon know which was which. I soon found that my mate was not going to be much help as, during his first week's employment, he had only learnt the two or three most obvious cows. He was the third Percy in this journal, his name being Percy Buck, always referred to as Puhsy. He was a first-rate stockman, and it was to him I owed a great deal later on, as he gave me no end of tips about matters I had not regarded as being particularly important in my earlier excursions into stockmanship. Things like knowing within an hour or two when a cow was going to calve, something which baffles quite experienced stockmen. This can waste a lot of time if there is a calf with the herd on a distant meadow at five o'clock in the morning, as there will always be a half a dozen would be mothers who will not be driven away from it, while the real mother may already be back at the cowshed, especially if she happens to be a young and inexperienced cow. Putting her in a box the night before is so much easier if you know what to look for. When it came to things like records, and learning the cow's names quickly, Puhsy was quite content for me to do it.

Three or four days into my first week, we received the monthly visit from the official recorder for National Milk Records. Bernard Bell was a big help with cow identification too. Some of the sticky labels had come off as the rain of our moving day had gone on and on, and he had to be sure as to who was who for his records. One way of doing this was to read the tattoo numbers in the right ear of each cow. At that time, the tattoos were put into the ears of the calves by the milk recorder as soon as possible after birth so that their identities were certain. Latterly it has been done by the cowman. The tattoo is a series of letters and numbers put in with a pair of pincers, the letters and numbers having spikes which go through the ear, and which are rendered legible by rubbing tattooing paste into the punch holes as soon as they have been made. The rubbing process is usually done using a convential toothbrush. Each series of letters and numbers is unique to that animal. As there had been many recent changes at Booters Hall, Bernard decided that it would be best to check all the tattoo numbers against those on the labels where these were still in place. Halfway through, he discovered a number that was not included on the list that he had on his record sheets. He would have to look it up with the aid of his books at home which contained details of all animals in the herd, or those that had left fairly recently. When we came to the end, he announced that we were one cow short. "Which one?" I enquired, knowing that I would scarcely be any the wiser when he told me, having only been with the herd four or five days. "Primula" he replied. Primula? I thought hard. Was she not nearly all black, with a zig zag lightning like edge to the black on her left flank? I could not remember seeing her that afternoon.

"Never mind! We'll find her in the morning I expect." Bernard said cheerfully.

When he arrived next morning, it was with the news that the mystery cow with the unidentified number was Do Do. I would have to make enquiries of the farm secretary, Miss Rutterford, as to who DoDo was, and how she came to be there. We milked. No Primula. During the morning, we searched all round the meadows that the cows were running on but could find no trace of her. That was on Wednesday. On Saturday, the feeders who put the sugar beet tops out for the cows, threw them out as usual on the day meadow, but also threw Sunday's supply onto a smaller meadow so that they had one job less next morning. On Monday morning, Puhsy went to get the cows up. We had already got into a routine. He got the cows up while I got the things ready in the dairy and parlour, and started milking as soon as the first arrived. In the moonlight, he discovered the little Jersey cow that was in the herd at that time- a Jersey sometimes being kept in a Friesian herd to boost the overall butterfat of the herd-standing near the hedge and peering into the ditch. He went across and shone his torch where the Jersey was looking, and there was poor old Primula on her back in the ditch and very dead. Presumably somebody had bundled her into the ditch during the Sunday night of the previous week, and two new cowmen who did not know the cows, had not missed her the next day. She was black, with a white lightning blaze on her left flank.

Still, we had found Do Do. Miss Rutterford looked through the records and declared that she had gone to market months before-but she had not! In fact, she was a good old cow and stayed with us for another two or three years. It never was worked out which cow actually went to market on the previous occasion. The result of having other cowmen who did not know the cows thoroughly either.

For a number of years, there had been a rapid succession of cowmen at Booters Hall. Some had stayed a year. Some had even stayed two, but equally, some had only stayed two or three months. I never could fathom out why. Certainly Mr Eglington was not a typical farmer, though he came from a farming family. He had his little idiosyncrasies. He was a member of the Plymouth Brethren, and if there was a special meeting at Toftwood Gospel Hall, he would invite everyone on the farm to attend, even helping with transport, though there was never any hint of compulsion. This was not really surprising, as he had spent many years as a missionary in Peru before he came back to take on Booters Hall and Church Farm as a single unit when arthritis became too much to tolerate in that country's torrid climate. I have always looked on him as the best boss I ever worked for. He paid me the top going rate for cowmen locally at that time, plus a bonus on milk produced and another for every calf reared to four days and disbudded. This was at a time when disbudding was done using a coating of thick caustic soda, thick in the sense that the consistency of the fluid was roughly like that of treacle. The horn buds were clipped and rubbed with a stiff brush sufficiently to erode the skin. The caustic was applied with the brush, just enough to cover the horn bud, and this dried on quite quickly with exposure to the air. After three or four weeks the resulting scab fell off and the animal was successfully polled. This method has been illegal for a number of years because no anasthetic was used. Today, dehorning is done at fourteen days plus using an anasthetic and a hot dehorning iron which burns the horn bud out under pressure. The animal does not feel the operation so far as we are able to tell, though even the best dentist is not 100% successful in suppressing pain! Having used both methods, I would be reluctant to say which is least traumatic for the calf.

In any given year when the various operations on the farm were particularly profitable, extra bonuses would be paid, not just to Puhsy and I, but to the ploughman, Cyril, or the spray man, Rupert, with proportionately less for other workers on the farm. During the best year that we had, one in which everyone on the farm shared, I personally received £50 on top of my guaranteed wage and regular bonuses. It may not sound much now, but as the best paid man on the farm I was getting £16 per week, so it represented the quivalent of another three week's wages. It was not all roses I will admit, partly because Mr Eglington's organisation could be chaotic, leading to unexpected problems where none should have been. He had a habit of remembering something as he was leaving, after a visit to the milking parlour, and I might catch a gem such as "Oh! Alec, the wassername needs seeing to." and off he would go. In time, I got used to everything being a 'wassername' and learnt to pin him down as to which 'wassername' he meant. I would know automatically to whom he was referring if he came into the parlour and said "How's old wassername this morning?" as it implied that she had been ill, but "Old wassername's bulling" had to be identified. Nonetheless, I am still puzzled as to why he had such a high turnover in the cowshed, and have always regarded our first five years at Cranworth as the best I knew, workwise. Even with a seven day start at five a.m. for the first two years, the work was not too arduous. Then came a reduction in farm worker's hours and we opted to take this in hours rather than extra cash and enjoy the luxury of a weekly lie in.

85 cows to milk twice a day in a four unit abreast parlour, muck out the parlour with a shovel and wheel barrow and wash it down. Wash up the milking and cooling equipment, still basically the same as that I had been used to at Debenham, the same milkline and releaser in the dairy and the same steam chest and sterilising system. One occasional source of trouble was the electrically driven water pump from a bore which pumped water into a relatively small header tank over the enginehouse, and which was supposed to cut in and out automatically. This only served the dairy itself, the cows being supplied from a bigger system installed at the main buildings about 200 yards away. Everything still had to be stripped down and have brushes pushed through all the pipework once a week, which was my job. Puhsy took great pride in keeping the parlour spotless. The walls were cement rendered and coated with Snowcem, a cement based whitening agent. Every splash of muck to arrive on the walls was religiously scrubbed off after every milking. The floor was clean enough to eat ones meals off and the collecting yard outside was as clean. During the winter, the feeding and littering of the covered yards in which the cows slept at night was done by two men off the farm, who also brought the concentrates from the tractor driven mill at Booters Hall. As I

mentioned a moment ago, the dairy unit was about 200 yards from the main farm buildings. This was a piece of typical Eglington logic. It was built there sometime after they took the tenancy of the farm, as it was near a junction in farm tracks and was roughly in the middle of the farm. As the plan was to have leys in the crop rotation which could be grazed by the cows, having the dairy unit in the middle of the farm would mean that when the leys were on the furthest fields, the distance that the cows would have to walk would be roughly the same in every direction. It is true that they did, and the hedges were still stock proof at that time, but the farm was something over 800 acres. After morning milking, if the cows were to graze a piece known as the 'Tin Sheds', we would have half an hour's walk to take them there, then walk back to the dairy, and then three quarters of a mile in the opposite direction to get home to breakfast. Looking back, I think that that was about the biggest grumble that I had. When grazing was close by the cowshed, we could make up for it as, when we had finished all there was to be done, we could go home until afternoon milking. On Sundays we were told to do just that which was necessary and then get home to out families. While we had four children, Puhsy and his wife had eight! The important thing was that both of us were conscientious, which meant that we never went home while there was anything that needed doing so everyone was happy.

One of the more important duties of the conscientious cowman is to spend some time each day just looking round his charges to see that all is well with them. At least, it used to be in those days, and I had sufficient time to do just that. By studying the cows regularly, I began to realise that they were not the dumb animals that scientists would have us believe. Since that time, I have given a lot of thought to the lowing herd from the viewpoint of the naturalist. The incident that triggered this train of thought probably stemmed from a cow called Cranworth Ben Jessica. Perhaps I had better explain that 'Cranworth' was the herd prefix, and that Jessica was the line name. Instead of having 'Jessica 1, 2 or 3' etc., a handy part of the bull's name was inserted between the herd prefix and the line name. In this case 'Ben' from 'Kelliways Benefit'. The idea was good when not carried too far, but when Mr E. decided that he would like to remember the grandsire by putting part of his name after that of the cow, resulting in'Cranworth Villa Jenny Rag' which is to say Jenny's daughter by'Lavenvilla Quack'and grand daughter of'Ragapple Casablanca' I felt that things had gone from the sublime to the ridiculous.

To return to Ben Jessica. The cows came into the parlour from the collecting yard, were milked, and let out through a gate which formed the front of each standing into a passage which led into the sleeping yard. To get out of the passage, they had to walk past the frontgates of the other standings, unless they had been in the standing nearest the outside door. All the gates had fitted feed bins in their bases, in which would be feed that was being eaten by the cows then milking. Thus there was scope for a greedy cow to do some stealing, which many of them did. Once outside the parlour, they were faced with a five barred gate which was closed, but as they pushed into the narrow end of the tiny triangular outside pen, their sides would come in contact with the gate which would swing open sufficiently to let them through, automatically closing by gravity after each cow had passed. Jessica came into the herd as a first calf heifer, as did all the cows as there were no bought in stock at all. I think that it was during her second lactation that she suddenly started putting in an appearance in the feed passage, stealing other cow's food when she should have been outside. At first, we just chased her out again, assuming that the gate had not closed properly behind her, but still she came back, so I stood and watched her. After the gate had closed, she turned round and walked up to it, put her chin on one of the bars near the head end and pressed down. Walking backwards, she took the gate with her, and when it had opened widely enough, she nipped smartly round the head end before it could swing shut again. Now it was our wits against hers. How could we stop her? Run after her shouting blue murder? Chase her with a stick and belabour her? We tried everything, and still she thought that the reward was worth the discomfort. Eventually I did come up with something that would deter her for the length of that milking. A quarter of a bucket of cold water hurled with all my force and aimed to land squarely on the white blaze in the middle of her forehead so that it splashed all over her front end, and she would go off shaking her head, and I have no doubt, cursing me under her breath! She had to be stopped though, as not only was she robbing others of their rightful and carefully calculated concentrate rations, she was also blocking the passage, and would not let anyone pass who was of lesser standing in the herd hierarchy. This may not seem particularly remarkable to anyone who has had experience of training animals, but I think it shows intelligence beyond normal animal training procedures in that she worked it all out for herself without any prompting from the herdsman. I shall have more to say about bovine intelligence later.

Another of Kelliways Benefit's daughters in the herd had problems of neurotic balance, if that is the correct term, similar to some which human mothers may encounter. This was Ben Countess, and a beautiful and aristocratic looking animal she was too, and the bane of our lives at milking time.

When a heifer calves for the first time, she has to be trained to be milked in the parlour, and is usually very nervous, what with the hum of the milking machine in the background and the pulsators ticking merrily away nearby, not to mention the indignity of having the cowman wash her udder and take the foremilk before putting the milking cups onto her teats, it is all very traumatic. The majority of the animals object more or less strongly to the milking clusters for a day or two. In order that they do not hurt either themselves or the cowmen, it is necessary to restrain them. Nowadays, this is usually done using a clever device known as a restraining bar, or 'kick bar' in common parlance. This is adjustable to the size of the cow. One end hooks into her flank, and the other goes over her nearside hip and hooks in front of the tailhead in such a way that, if she tries to kick, the top end of the bar presses into a nerve and prevents her from doing so. When we first went to Cranworth, the kick bar was either still in its infancy, or had not even been invented, so we used a leg strap. This was a broad leather strap about three inches wide which was fixed round one leg above the hock, through a slide, and pulled up tight round the other leg with a buckle. Most heifers got used to being milked within a couple of weeks and the leg strap could be left off. Not Countess. She had to be strapped for every milking all the way through her first lactation. When she calved for the second time, she came in as quiet as a lamb. We milked her with no strap and everything in the garden was lovely- we thought. This went on for six or eight weeks and it came to be time to serve her again. Suddenly, she was back to her old tricks, and it was back to the leg strap for every milking for the rest of her lactation. Though we did not do pregnancy testing at Booters Hall, relying on our own judgement as herdsmen had since the beginning of time, we knew exactly when Countess became pregnant again each lactation, as that was the day when we had to start using the leg strap again.

I mentioned pregnancy testing and said that we did not do it. Very few farms did at that time and, provided good records were kept, such refinements were not necessary. The number of cows that were assumed to be in calf, and which turned out not to be was certainly no higher than 1-2%. Even where pregnancy testing is carried out, the failure rate can be as high. The test is normally done 6-8 weeks after service, and a few cows will reabsorb the foetus between 8-12 weeks and no one any the wiser. If a cow failed to come on heat we would get the vet to check that her ovaries were working after about 10 weeks, but most other things we had to do ourselves at Booters Hall. If a cow retained her afterbirth, or 'cleansings' as they were known 'in the trade', I had to remove them myself, usually at about six days after calving. This was a particularly smelly job, as by that time the offending material would be getting a bit 'high', and we did not have the benefit of a throwaway plastic sleeve-cum-glove in those days. It was a case of wash as thoroughly as possible, rub your hand and arm with Super Kinray(without taking the skin off) and then washing again and hoping that you would not take the smell home with you. Pessaries had to be inserted when the removal had been effected, and a couple of weeks later, the cow had to be 'washed out'. Nowadays, an antibiotic is inserted into the uterus using a catheter. No antibiotics for that operation then. Mix up a weak solution to the volume of half a gallon with dettol and warm water. Suck half of it up into a huge old brass syringe with a flexible nozzle on the end, and a 'pump handle' on one side in the manner of a large grease gun. Hold the syringe with one hand and take the nozzle into the vagina with the other, carefully inserting into the cervix. Close the pump handle smartly, propelling the contents of the syringe into the uterus. Remove, refill the syringe and repeat the operation. Did they ever breed again after that sort of treatment? To tell you the truth, so far as I can remember, they bred as readily as they do today using the much more sophisticated antibiotics administered by a qualified vet. Of course, Mr Eglington, having been a missionary operating in a part of the world where medical and veterinary care were pretty minimal, had learnt to do a lot of things for himself that most 'lay farmers' would not have dreamed of doing. On one occasion, we had a cow that was obviously dying, even in spite of having been treated by the vet, and she was given a minute injection of strychnine by Mr E. to stimulate her heart, and perhaps give her a little longer to respond to the vets treatment. She rallied for a short while but eventually succumbed. The important thing to my mind was that the knowledge was there and it might have worked.

There was one area of veterinary care where my views and his did diverge somewhat, and that was with regard to calving. All the calving boxes were in the yard at Booters Hall so, during the first four days after calving, the cows had to be taken back at night to their calves to be suckled. This also meant that the calving was all done at Booters, and that Mr E. himself was in charge. Admittedly, I did not have to keep running to and fro to see how they were getting on as this was his prerogative. He was always very anxious to spare the calving cow unnecessary suffering, and would seek to 'help' them as soon as possible. In other words, as soon as the calf's front feet were in view, calving lines would be attached, and theoretically, a steady gentle pull would do the trick. All too often, this would not be the case and, at one, or two, or three o'clock in the morning, some gravel would rattle against my bedroom window and it would be Mick Clayton, our neighbour and "Could you go over to Booters ? There is a cow calving and Mr Eglington wants some help." There would follow perhaps one, perhaps two hours of often unnecessarily hard pulling before the offending calf saw the light of day. I became quite convinced that the desire to help actually increased the trauma and suffering for the cow. I became even more convinced of this by the time I retired nearly thirty years later that this was so, having seen the heavy handed approach of too much pulling too soon which has resulted in a cow being off her feet for days or even longer because of nerve damage.

My own policy for many years was to go away for half an hour and do something else, when the cow had got to the stage of having the calf's front feet visible. Often, when she gets to that stage she will have quite a long rest and may not engage in 'straining' or labour, for some considerable while. If, after half an hour, she has obviously been straining and getting nowhere, I will then lend a hand, but if she is wandering round the box restlessly, or if she is obviously making steady progress, then I will go away again, perhaps for ten minutes or so. If she had been progressing, by this time she will probably have the calf and will be licking it. If she has been wandering around, she may by now have laid down and be ready to get on with the job. That was how I preferred to act in 95 cases out of every 100. The other five cases were likely to have been abnormalities such as breech calves, or those that have arrived at the way out somehow in the wrong position. A breech calf is immediately obvious as one first sees the backs of the feet instead of the fronts. Even then it is advisable to give the cow sufficient time, if she is making progress, to get the feet out almost to the hock before intervening as this will mean that the cervix has opened sufficiently for the calf to be removed with a minimum of force. As soon as pulling starts and progress is being made, and this is not as easy as with a normal birth because the 'mid-wife' is having to pull against the lay of the calf's hair, the breech calf must reach the open air in the shortest possible time as the navel cord will break at a fairly early stage, the calf will start breathing and will fairly quickly drown in the fluid within the placenta. If the calf is removed quickly, no further action may be necessary, but if it has had rather a slow passage it may be necessary to hold it up by its back legs so that any inhaled fluid can be drained from its lungs. Many breech calves seem to be abnormally large, and in such a case, it may be necessary to haul it up over a half door or partition of some kind in order to get it high enough. While this may sound drastic, one should remember that a new born Friesian calf can weigh as much as a cwt, or roughly 50 kg.

In the small number of cases in which the presentation is not straight forward, or backward as the case may be, it soon becomes obvious that something is wrong , and the herdsman will have to make an internal examination. It may be no more than a foot doubled back, in which case it will be necessary to push the calf back somewhat in order to raise the offending foot to the correct position. Not as easy as it may sound either as, while you are trying to push the calf back, the cow will redouble her efforts to push it forward. About fifty per cent of wrong presentations can be sorted out without recourse to veterinary assistance. The rest, perhaps two or three in the course of a year in a large herd, really do need the help of an expert. Actually getting a badly displaced calf to the correct position for birth requires a good deal of strength and endurance on the part of both the vet and the cow. Just twice in a lifetime, it has proved impossible to calve a cow by normal means. The first occasion was soon after I left school and was looking after the cows at home. The vet, with very considerable help, failed completely to make any progress and in the end, we had to call the knacker to come and put the poor animal down. When he cut her up, he found a badly deformed calf with two heads, seven legs and eight feet, whilst all the vet could feel from examining her was a normal calf.

The second case, as recently as during the 1980s, involved a presentation of four feet which were at first thought to be twins. However, when one set of legs was pushed back, the other pair went with them. On this occasion, the vet correctly diagnosed an 'inside out calf' or schistozome. By this time, it was considered feasable to do a caesarian operation on a cow, and that is what he did. Yes, he was able to give the cow an anasthetic, but no, he had not got a nice clean operating theatre with spotlessly white walls and masked theatre staff. I did find a piece of plastic sheet to put all the bits and pieces of equipment on, and the operation was done in the ordinary straw filled calving box. The cow recovered very well after her operation and went on to have one more calf after that.

Ever after 1963, I had a quiet dread of the fourth week in January.

Have you ever noticed how the winter seems to drag at about that time? Starting work at five in the morning and finishing at about six in the evening means that, by the end of January, you have been starting and finishing the days work in darkness for some ten weeks, ever since the clocks changed at the end of October. Toward the end of January, it is possible to see remnants of the afterglow still by the time evening milking is complete. Then bang! There comes a week or more of savage frosts and the coldest nights of the winter at some time between the 20th and 30th of the month, and you begin to feel that it will never end. The 1963 freeze up actually started on Boxing Day 1962, and for ten long weeks we had frost and snow. Puhsy and I were both old birds not likely to be caught on chaff, so we kept one tap trickling slowly into the cow's water tank 24 hours a day, and thus avoided a complete freeze up round the parlour. I think it was the 25th when Puhsy had his day off that week. Peter Page, the then farm lorry driver was doing the relief milking, and the forecast had been for one of the worst frosts of the winter. The milk was released from the recording jars through big brass master taps which had holes going through the core of the tap in different directions to correspond with a variety of operations. As these were susceptable to frost, I had taken them out overnight. In addition to being in the middle of the farm, the dairy unit was also close to the Blackwater River, the infant River Yare, so was in about the coldest position possible. Also, heaters in cowsheds were an unheard of luxury in 1963. We had always made a habit of wetting the floor first thing in the morning, before any cows came in, to make it easier to get the muck off. When we got to work, I told Peter not to put any water down or we would have had a skating rink in a very few minutes. I got a bucket of water from the boiler, kept banked up overnight and everynight as I had learnt from Putt, dipped the master taps into the nearly boiling water and inserted them into their housings, and got everything else ready while Peter was getting the cows into the collecting yard. I went round to start the motor that drives the vacuum pump and pressed the switch. 'Clunk'! Nothing happened. The vac-pump was frozen. I did not like hot water and electricity together, but there was nothing for it but to pour some hot water over the vac-pump. Bingo! It started first time. I went back into the parlour. The master taps were all frozen solid in their housings. When they had all been thawed out again, we were able to start milking. We were both clad in two pairs of trousers, three pullovers, our jackets, and government surplus overcoats which we normally only wore outdoors, and were absolutely frozen. The cows stood outside in

Moor Lane, Cranworth, little used by 1978

the snow, patiently awaiting their turn to come in, and where each stood, she melted a little snow with the warmth of her hooves. When she came in, her feet were damp and by the time we had finished milking, we actually had a layer of ice half an inch thick all over the parlour floor. The nearest maximum/minimum thermometer to our dairy unit registered -4° Fahrenheit which is about -20° Celsius, and down near the stream it was likely to have been even lower.

Nor was this the end of our problems. Moor Lane, which ran from the road to the cowshed, was three quarters of a mile long, and halfway down the lane, a spring bubbled up from somewhere, actually running down the track for some distance before finding its way into a ditch on the opposite side. Moreover, it flowed unabated throughout the freeze, the water running under the large sheet of ice that covered the lane to a depth of two or three inches for 20 yards or more. The churn lorry could undoubtedly have got down the lane, but would have found it impossible to get up again with an extra thirty churns full of milk. Leslie was given the job of collecting the churns each day on a four wheel trailer behind the tractor and taking them up and parking them at the top of the lane near the road. Most of the the time, he did the job without any problems, but on one never to be forgotten occasion, he turned too sharply at the top of the lane and overturned the trailer. He leapt off the tractor as quickly as he could and grabbed up those churns which had not lost their lids, but even so, he added something like 180 gallons of milk to the ice.

There was one thing abut Eric Eglington when faced with an incident of that nature, He never raised his voice, or made a show of laying blame. It was always "Oh well, accidents will happen sometimes." I have a feeling that that attitude went a long way toward making his workers anxious to see that accidents were few and far between.

As I have said, the feeding and littering of the herd was done by some of the other men on the farm, and it was from their ranks that the relief milkers were taken, though most were not very keen on the early mornings. One of these was a relatively younger member of the farm staff, Terry Smith, who had never done the job before. He was inclined to be of uncertain temperament, and if things went wrong, he would soon start shouting and swearing at the top of his voice, being audible from some distance beyond the confines of the cowshed. I taught him as well as I could and he became the regular relief milker towards the end of our stay. Eventually, he married Puhsy's eldest daughter and they had two sons, but separated after some twenty years as is the unfortunate way of the world today.

Puhsy was of fairly even temperament, though even he had his moments. Away from the cowshed, he tilled his large garden assiduously, growing mainly greens and potatoes. With the latter, he had an annual target of $1\frac{1}{4}$ tons, which was sufficient to last his family until spuds came round again, and I think he probably managed enough cabbages, sprouts and cauliflowers to keep the family in greens as well. During the long dark winter evenings, he would lay on the couch from tea time until bedtime, watching the telly, and he would be in bed soon after nine every night. Remember that we started work at five. Puhsy was up most mornings soon after four, without the aid of an alarm clock, and would have a shave before he came to work. Both of us had young families, and when Christmas came round, Mr Eglington told us, "Start as early as you like Christmas morning and get home to your families as soon as you can." I talked it over with Puhsy and we agreed to start at half past three, which would mean that we should be finished and home by eight. When I arrived at the cowshed at half past three, the milking machine was already humming merrily away and several cows were already back in their sleeping quarters. Puhsy had woken up just after two and had been at work by three! As we made our way home, we met the feeders, just starting work, and we both got home in time to see the stockings unpacked.

Puhsy was not so good in fog though. One very foggy October morning the cows had to be got up from a forty acre ley about a quarter mile away. Off he set with his torch while I got things ready and prepared to start milking. After a very long while, a handful of cows arrived , followed at lengthy intervals by one or two more. A flurry of activity brought the total arrivals to just over twenty, with Puhsy behind them.

"Where are the rest then?" I enquired.

"Out in the middle o' the fild somewhere I 'spect. I went all the way round the hedge an' brought all those I found, but I wern't a goin' away from the hedge. I wouldn't go no further away from the hedge than me light ud show. I done that once during the war when I was in the hoom guard an' I lost meself, an' I shan't do that again in a hurry."

By now, the fog was beginning to show the first glimmer of daylight, so Puhsy set off once more, to return quite soon with the rest of the cows, most of whom had heard him the first time and were following along the lane in the wake of their fellows.

An occasion when Puhsy did lose his temper was when his wife won £60 on the weekly Labour Tote, which was a regular lottery run by the local Labour party. Sixty pounds in those days was the equivalent of a month's wages, so it was not to be sneezed at. But the family were celebrating noisily downstairs at about half past nine in the evening, and Puhsy was upstairs trying to sleep. Suddenly, he could stand it no longer. He came downstairs in his nightshirt like an avenging angel, grabbed a broom and drove the whole family upstairs to bed.

If he had a failing, it was the ability to laugh uproarously at other people's misfortunes. I was on the wrong end of that on a number of occasions, though only minor happenings. He would unfortunately laugh when the other party was injured. For instance, on one occasion when Terry was relief milking, he came through from the dairy incautiously, stood on a slippery patch on the floor and his feet went from under him. On landing, he caught the base of his spine on the dairy doorstep which, though no lasting damage was done, hurt considerably at the time. Puhsy went off into peals of laughter. Just beside Terry was a bucket and the tank of water used for washing down. Terry grabbed the bucket, dived it into the tank and hurled the contents over Puhsy in one movement.

"Now laugh that off you silly old bugger!" Puhsy went home and changed.

One of the regular feeding team was Bob, Leslie's father. Where Leslie was tall and well built, Bob was small and thin featured. Leslie took after his mother. Bob would come in and have a word after he had finished feeding on a winter's afternoon, and we heard quite a number of amusing anecdotes from him, though perhaps the most amusing was told, not by him, but by me about him.

He came in one afternoon, and it must have been quite mild as he was in his shirt sleeves. Moreover, he was pulling one sleeve round and at the same time craning his neck in an effort to see what was on the back of it.

"Alec! Is there a big lump of muck on the back of my shutsleeve?"

I went over and had a look.

"Well, I don't know about a big lump Bob, but there's certainly a lump there. Here, wait a minute and I'll get my knife and scrape it off."

Half a minute with my knife blade and only the stain showed where the offending dollop had been.

"There Bob! That's all off now bar a little stain." Bob was not convinced and kept trying to twist his sleeve round to see how big the stain really was. I tried to reasssure him again.

"Yes but-" he wailed, "whatever is my missus a-goin' to say when I hop into bed tonight?"

I'd better let Puhsy have the last word here. One of the most excruciting pains one can get in the cowshed, is obtained by having a cow stand on ones foot, especially if she manages to turn her foot at the same time, giving the pressure a twisting action. On one occasion, I heard a wail of anguish from Puhsy and looked round to see him hopping around on one foot while he clutched the other with both hands.

"What's the matter Puhsy?" I enquired innocently.

"Blast bor! I're found a jamtoes nest!" came the reply.

Looking back, it does not seem possible that things could have changed so quickly. Colin Holmes lived with his widowed mother in the cottage beside the Church Green and worked on the farm. Church farmhouse housed the Claytons and Bulls. The cottages opposite our drive held workers in either end, including Cyril Kerry who did all the ploughing on the 800 acres. Using a Fordson Major and either two or three furrows according to conditions at any given time, Cyril would be out on his tractor from ten past seven every morning, often until 9.30 at night, five nights a week and until four oclock on Saturdays, from the end of harvest until late January or early February, when he would finish ploughing the land that had been sugar beet. Leslie and his brother in law, George Everett, always known as'Plonky' lived in a pair of cottages beside the farmyard. Puhsy and his family lived in a detached cottage a little way down Moor Lane, whilst on the council housing estate lived Bob Williamson, Rupert Newson who did all the spraying and other tractor work and Cyril Spragg, while Terry Smith lived with his parents. In addition, Langer Webster cycled the three miles daily from Scoulton to look after the pigs and young stock, and

Billy Boast cycled from Shipdham until he got the sack and was replaced by the Mortlocks, father and son, from Reymerstone. This amounted to twelve or thirteen men all working on the farm in the sixties, ten or eleven of whom were Cranworth residents, most being of long standing.

Why mention this? Because it is relevant to the cataclysmic change in the life of both farm and village that came about between the 1960s and 1990s, and has been echoed and re-echoed in many villages up and down the country, and nowhere moreso than in the highly intensive farming areas like East Anglia.

In 1960, Eric Eglington was farming his 800 odd acres, rented from the East Anglian Real Property Company which at that time was a Dutch owned consortium. Based at Cantley it had been set up at the same time as Dutch money and expertise had been used in the setting up of the first commercially viable sugar factory at Cantley in 1912. There had been an attempt half a century earlier, when a small sugar factory was built at Lavenham in Suffolk, but this only operated for a few years in the 1850s. In 1960, E.A.R.P. owned and rented out or farmed some 20,000 acres dotted all over Norfolk, with the majority of the farms 'in hand'. Eric was farming in what was the transition period between 'traditional' and 'modern' farming methods and, though he used no horse power, neither did he have sufficient tractor power to keep absolutely on top of everything on the farm. Consequently, he had had to let some things drift. Keeping a relatively small staff to farm 800 acres by 'neotraditional' methods, he had little hope of covering areas which at that time still needed hand labour. Such things as hedge trimming and cleaning out ditches. By 1965, he had been there about 25 years, and hardly a hedge had been cut or a ditch cleaned out in that time.

Meanwhile, during the winter of 1961/62, the British Trust for Ornithology issued an appeal for volunteers anywhere in Britain to take part in a new type of survey. This was to be called the Common Bird Census, hereafter referred to as the CBC, and was being funded by the Nature Conservancy, as it was then known, to try and monitor the effects of toxic pesticides, and especially seed dressings used on autumn sown crops which at that time contained Aldrin, Dieldrin and Heptachlor. As most people are aware, these dressings decimated populations of bird species at the top of the food chain, as well as those which actually ate the poisoned seed. Thus the seed corn may have been eaten by rats, mice, voles, sparrows and other seed-eating birds including some larger species like Pigeons and Rooks. These were poisoned, though not all died, and those that did not were in turn caught and eaten by Owls, Kestrels, Peregrines, Sparrowhawks and other predators. Some of the latter were to eventually die from the build up of toxic elements in their bodies, but more importantly, those that did not die, were rendered more or less sterile, or produced soft shelled eggs which the birds could not successfully incubate. All predator species then suffered a disastrous population crash. Hence the need for the special study.

I volunteered and took under my wing 246 aces of the Booters Hall/Church Farm complex, which I had to walk round ten times between approximately March 21st and June 21st, listing every bird seen or heard during a 4 to 5 hour stroll. 'Listing' is not really the correct word, as each bird had to be entered up on a visit map of the area. This was taken from the Ordnance Survey 25 inches to the mile series, of which, every farm has a copy relevant to their property. A simplified photocopy of this map is the basis for all CBCs. Having done my 40-50 hours per year doing the field work, all the information had to be transferred to species maps with, on average, four species being entered on each map, each specie's records being in a different colour. As there were anything up to 60 species breeding on the plot in the early days of the survey, plus quite a number 'recorded' but not breeding, it is easy to understand that the whole operation took up a considerable amount of spare time each year, especially after we moved to another herd 14 miles away, and I managed to get permission to come back and carry on with the CBC. In fact, I continued until 1990, a total of 29 years unbroken recording, only ending when the doctor warned me to take things a bit easy when I started to develop asthma. Why did the CBC have to be the thing to go? The main reason was because it had to be done during the first hours of daylight, when the birds were giving time to singing and sitting about prominently on territorial perches, whilst later in the day they tend to be much less visible. Thus, during the latter part of the recording season, I would actually have to be over at Cranworth on my day off to start my 'plod' at an earlier hour than I would have been starting to milk back at home.

During the 29 years, many things happened and were monitored, though the original reason behind starting the CBC failed to materialise on my plot. The first summer that I carried out the census, 1962, there were 20 pairs of Wrens resident on the 246 acres. The following winter I have already described, with its ten weeks of savage weather. The 1963 Wren population consisted of a single pair and

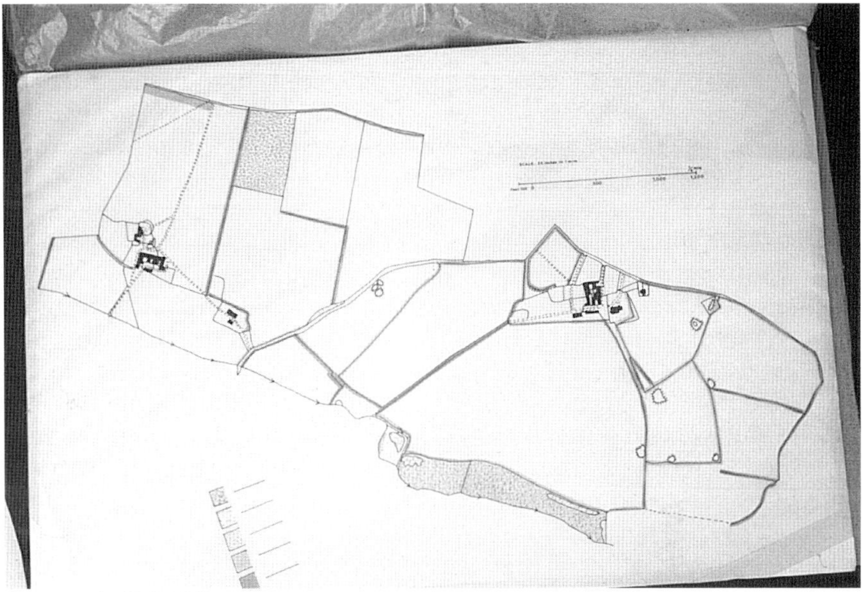

Cranworth CBC 1962

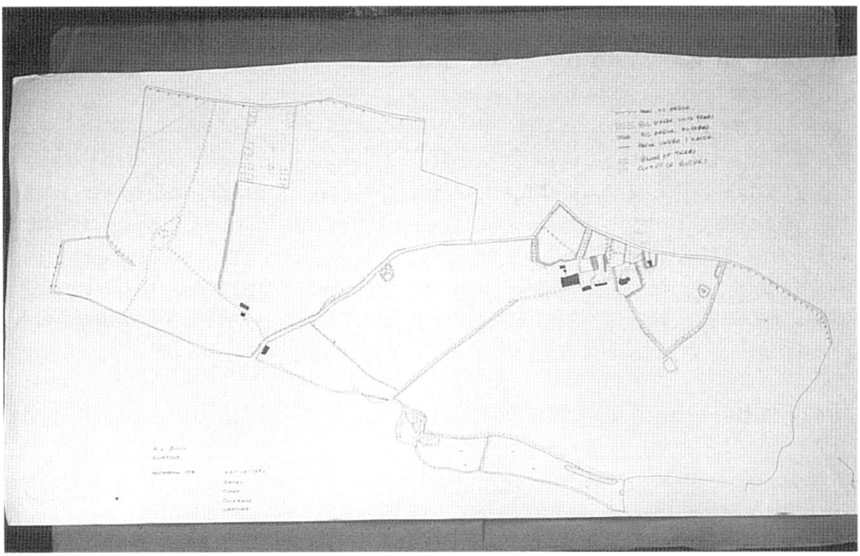

Cranworth CBC 1978

three wandering males which did not breed because they could not find a mate. This state of affairs was mirrored more or less by almost every resident species, and it is interesting to discover that, three relatively mild winters later, most species had recovered to their 1962 levels. Other things threatened 'my' birds though.

EARP wanted to get Mr Eglington out and take the farm in hand. The only way that this could be done at that time was to find him guilty of bad husbandry and give him notice to quit. At Michaelmas in 1965, they gave him twelve months notice to cut all the hedges and clean out all the ditches on the 800 acres or he would get his marching orders. Contractors with hedge cutters and ditch diggers were brought in, and all through the winter, spring and summer following, the screech of the circular saw mounted on a telescopic arm type hedgecutter rent the air day after day to a background noise of steadily throbbing JCB engines and the monotonous unoiled squeaking as their soil laden buckets deposited a generation of spoil from the ditches onto newly cleared hedgebanks. Things got so desparate that the operations were carried on in the ripening corn, right up to harvest, but the deadline was met, and things might have subsided back to normality.

Bird populations did not immediately suffer, though the trauma they underwent during that spring and summer was dreadful to behold. As they built nests in hedges and bush clumps, so were they torn out from under them and they had to start again in the piles of debris, only to see the debris heaped up into vast bonfires.

By now, Mr Eglington's eldest son had returned from a three year course at Cirencester Agricultural College with new ideas and new ways of doing things. Some of the ideas were alright, others not. He wanted to run things his way and I did not feel like having my own position eroded which resulted in a personality clash and I handed in my month's notice. Mr E. tried to persuade both of us to reconsider, but it was too late, and we left. Puhsy, who had started the week before I did, seven years before, left the same week. About a year later, Mr Eglington collapsed and died of a heart attack, so EARP got their desire, as they could take the farm in hand by not allowing the tenancy to pass from father to son. A farm manager was put in and a vast 'rationalisation' programme took place over the next-few years, mainly between 1970 and 1972. In other words, the guts were ripped out of the farm and nothing could ever be the same again.

Remember, my plot, at 246 acres, was just under one third of the total farm acreage. On my plot alone during 1971 and 1972, two and two thirds miles of hedges were bulldozed out and their associated ditches filled in. Every internal hedge on the plot except one disappeared forever. The one to escape was one which had been about the last to get grant aid when Eglingtons had it deepened and widened

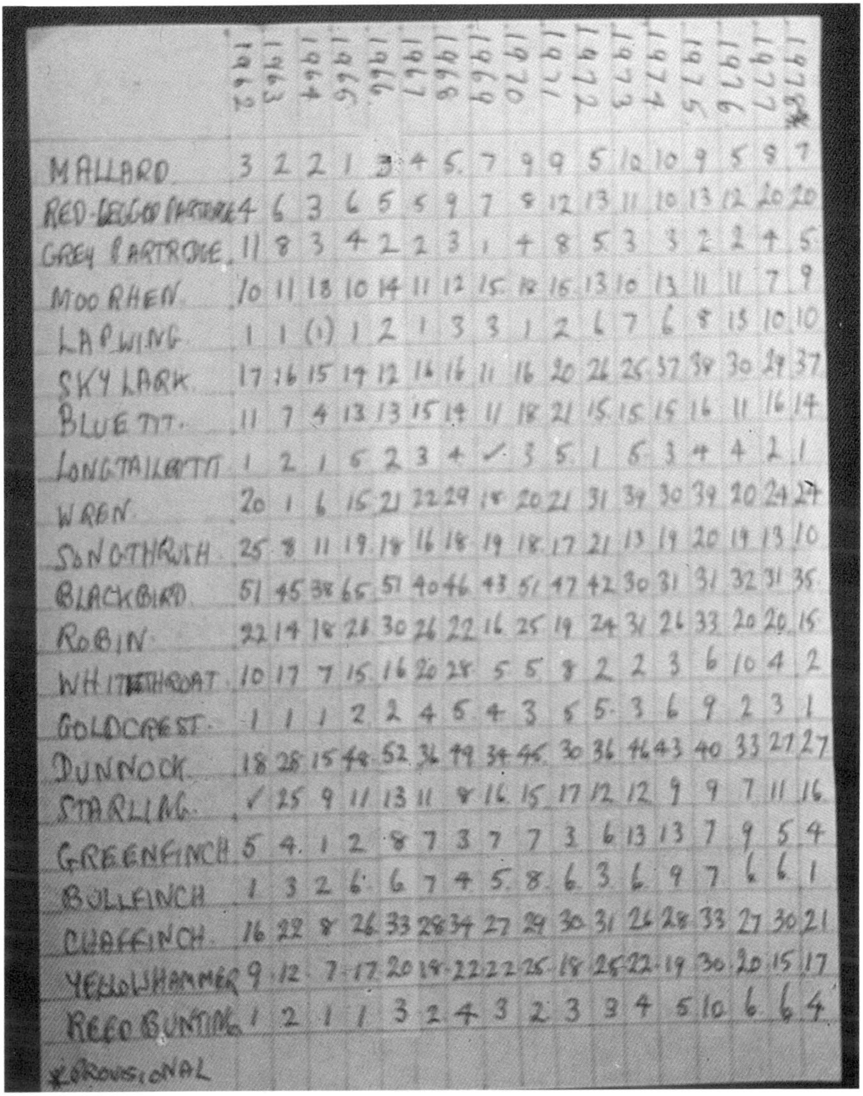

	1962	1963	1964	1965	1966	1967	1968	1969	1970	1971	1972	1973	1974	1975	1976	1977	1978*
MALLARD	3	2	2	1	3	4	5	7	9	9	5	10	10	9	5	9	7
RED-LEGGED PARTRIDGE	4	6	3	6	5	5	9	7	8	12	13	11	10	13	12	20	20
GREY PARTRIDGE	11	8	3	4	2	2	3	1	4	8	5	3	3	2	2	4	5
MOORHEN	10	11	18	10	14	11	12	15	18	16	13	10	13	11	11	7	9
LAPWING	1	1	(1)	1	2	1	3	3	1	2	6	7	6	8	15	10	10
SKYLARK	17	16	15	14	12	16	16	11	16	20	26	26	37	38	30	24	37
BLUE TIT	11	7	4	13	13	15	14	11	18	21	15	15	15	16	11	16	14
LONGTAILED TIT	1	2	1	6	2	3	4	✓	3	5	1	6	3	4	4	2	1
WREN	20	1	6	15	21	22	29	18	20	21	31	34	30	34	20	24	27
SONGTHRUSH	25	8	11	19	18	16	18	19	18	17	21	13	19	20	19	13	10
BLACKBIRD	51	45	38	65	51	40	46	43	51	47	42	30	31	31	32	31	35
ROBIN	22	14	18	28	30	26	22	16	25	19	24	31	26	33	20	20	15
WHITETHROAT	10	17	7	15	16	20	28	5	5	8	2	2	3	6	10	4	2
GOLDCREST	1	1	1	2	2	4	5	4	3	5	5	3	6	9	2	3	1
DUNNOCK	18	28	15	44	52	36	49	34	46	30	36	46	43	40	33	27	27
STARLING	✓	25	9	11	13	11	8	16	15	17	12	12	9	9	7	11	16
GREENFINCH	5	4	1	2	8	7	3	7	7	3	6	13	13	7	9	5	4
BULLFINCH	1	3	2	6	6	7	4	5	8	6	3	6	9	7	6	6	1
CHAFFINCH	16	22	8	26	33	28	34	27	24	30	31	26	28	33	27	30	21
YELLOWHAMMER	9	12	7	17	20	18	22	22	26	18	28	22	19	30	20	15	17
REED BUNTING	1	2	1	1	3	2	4	3	2	3	3	4	5	10	6	6	4

*PROVISIONAL

Cranworth CBC Summary 1962-78

and, for some reason, the Ministry baulked at paying another lot of grant aid to fill that particular ditch in again. As luck would have it, it was about the poorest hedge on the plot for bird territories, consisting almost entirely of elm and, as the hedgerow removal coincided with the arrival in Cranworth of the virulent strain of Dutch Elm disease then sweeping through Norfolk, most of the hedge promptly died anyway. The decline in the populations of hedgerow nesting birds was not as immediately dramatic as I had expected, though it was inexorable and largely irreversable. Admittedly such species as Wrens did, over a period of mild winters in the late 70s, manage to spread out along entirely unsuitable 'cow parsley hedges', nesting for a time in vegetation rather than bushes, but a sharp winter in 1979 soon decimated their number again and they were much slower to recover than after the 1963 winter.

Admittedly some ground nesting species such as Skylarks did benefit from the 'Big Field'. From a pre-clearance population of about 16 pairs, they rocketed to three times that level when the recently created 83 acre prairie had a thin crop of Italian ryegrass for seed, and then fell back to an annual average of 25-30 pairs.

What else happened on the farm? As some of the older workers reached retiring age they were not replaced, and so the number employed gradually dwindled. When they took the farm in hand, EARP scrapped the dairy unit I had worked in and built a completely new one twice as far from the road as the old one. To my surprise, Terry was taken on as cowman, though after a year, when the new herd size doubled with a new influx of heifers, he preferred to carry on as second cowman as he was not too keen on the amount of paperwork necessarily done by the head man. After a number of years the Dutchmen sold out. I seem to remember that the figure for the whole estate back in the late '70s was something like £5,000,000, to the Coal Board Pension Fund. In practice, this made no difference at all. The name stayed the same and the workers stayed the same, though there was a change of manager at one stage which resulted in an even more rigorous elimination of odd corners which had been of value to the remaining wildlife.

Strangely enough, the final chapter in the tale of a farm and parish was precipitated by the Common Agricultural Policy and the onset of milk quotas. In the early days, it was not possible to buy and sell quota as it is today, and the quota system meant that milk producers had to produce a good deal less than they had been doing. However, it was possible to transfer quota from one farm on an estate to another. Somehere about 1985, Leslie, who had worked on the farm since he had left school, was awarded his long service medal at the Royal Norfolk Show for 41 years continuous service on the same farm. The next week end, the bombshell landed. All the men on the farm with the exception of the two cowmen received their redundancy notices. The cowmen were to carry on until the following spring, by which time a vast new dairy unit would have been completed at Reedham, near Cantley, and three herds on the estate were to be made into two, the Cranworth herd dividing its quota between the other two which would then be able to continue production at maximum capacity. The following spring, Terry and his mate saw the cows off to Reedham, Terry was redundant and the head cowman was able to take early retirement.

After some months, Terry got a job with a firm, travelling round East Anglia catching chickens in deep litter houses ready to be processed, from where he moved to actually working in the processing factory. Some years later, an accident at work left him partially disabled and unable to work again. Leslie, 55 at the time and diabetic, was also unable to get further work.

For the next two years, the Cranworth farm was run and farmed from the next nearest EARP farm at Sporle, 12 miles away. No one locally was now employed on the farm at all. There followed the final upheaval. As agriculture became less profitable, the pension funds started pulling out, and the Coal Board were no exception. The whole of EARP was sold, but not as a going concern. This time, each farm was put up separately and sold to whoever would buy it. In most cases, this was a nearby farmer. At Cranworth, the farms were divided, the Booters Hall part being bought by Letton Estates which were the property of and managed by a great nephew of Eric Eglington. A new house has built on the site of the former house where Eric and his family had lived, and a partner in the business now lives there. The Church Farm part of the estate was bought by a consortlum that farms at Scoulton, 3 miles away. In both cases, efficient arable farming is the order of the day, a state of affairs which is unhelpful to wildlife generally, as of course, with no one living and working locally on either farm there is no room for livestock, and basically where livestock is kept there is always much more scope for wildlife to prosper, especially birds. About the only plants other than grass which can persist around the fields are those with some resistance to sprays such as Cow Parsley, Hogweed and Mugwort. Long gone are the Primrose and Violet studded banks round the fields in spring, to be followed later by such species as the lovely deep blue Wild Columbine which was quite common in the district as a whole when we first moved to Cranworth. Now, I know of only two sites where I feel confident that the plants are the true native Columbine. One on a roadside bank adjoining a little old meadow near Reymerstone, and the other on a private nature reserve at Carbrooke. If no one is now employed on the farms from the local population, what has happened to the houses? Why did the new owners have to build a new house at Booters? What happened to the rambling old farmhouse where Eric Eglington and his family lived? It had stood on the site, at least in part, since the time of the first Elizabeth, with other parts added on more recently. When the farm was taken in hand, it was decided that the house and buildings at Booters were surplus to requirements. The buildings were quite an extensive range and in reasonable condition, but they did not lend themselves to large modern machinery or modern methods, so they were bulldozed and the site ploughed out. Being away from the road and nearer the centre of the farm, the house was thought to be a liability. Somebody knew somebody, and the old house was burnt to the ground as part of a film sequence for one of the television 'Tales of the Unexpected'.

Church farmhouse had been put back into one and modernised for the EARP farm manager. This was repaired and modernised further, the nearby barn was also converted into a 'des-res' and both are now among the more desirable properties in the village. Terry still rents and lives in his cottage opposite Church Farmhouse, but almost everybody else has passed on or moved away and where they were farm cottages, most have been sold and are now privately owned. The school was closed sometime during the 1980s and is also now a private house. Virtually no one in the village now has any connection with the land. All have to commute somewhere else to work and during the day, the village is dead, apart from the inhabitants of the old people's bungalows on the housing estate, a few of whom may still have memories of a tight little community not so very many years ago.

Foxley

Our next stop was with Ted Littlewood, a genial farmer of Suffolk stock who was about my age, had cows because they paid, did his own relief milking but otherwise was not particularly fond of them, and was content for me to run things my way. They were milked through an abreast parlour such as I had been used to, and yarded at night during the winter, just going out during the day for kale or beet tops, the feeding being done by the other men on the farm. I had a few extra duties here, being within the farm buildings instead of a quarter mile away, one of which was the grinding. Not just the concentrates for the cows or young stock, but also sufficient of a different ration to feed the pigs, the province of the pigman-cum-stockman, George Hall who also looked after the young stock once I had reared them to eight weeks. We lived in a detached house opposite the little green with the Foxley war memorial in the middle beside the then A1067, Norwich-Fakenham road, though now on the road to Foxley West. George also had a smallholding down the lane past our house, with two or three acres of land and his own piggery which he managed in his spare time.

The number of milking cows was between fifty and sixty. A few were Friesians bought at sales from other herds, some even bearing pedigree names. Some were Ayshire crosses and there may have been other strains as well. In addition, Ted had a penchant for keeping a few suckler cows round the place. One was a Hereford and the other three were Friesian x Charollais. Herein lay a hidden hazard as from time to time, their calves would be deemed to have suckled long enough and would go to market. For a few days, or for a week or two, until more calves became available, they would be my most unwelcome guests. The three crosses milked reasonably well, no doubt from the Friesian in their ancestry, but when it came to being milked, all four had the strongest possible objections to the arrangement and I had another insight into cow psychology! The only way to manage what proved to be a very hazardous undertaking was a leg strap on each.

We also had another deviation from the norm during my first years there, in that we had a Friesian bull. I have no doubt that he was a very good bull. Unfortunately he had been bought unregistered just at the moment when the regulations relating to bulls had 'suddenly' been changed and he became illegal. Ted kept him and used him and, when the vet came to do the annual tuberculin test, he put the bull down on the form as a 'steer'. However, by the third year, he said, "Look here Mr Littlewood, when I come to do the TB test this year, I am NOT registering that old bull as a steer again. "So he had to go, much to my relief as he was not of the nicest of temperaments.

However, by that time I had persuaded Ted that it was in his own best interest to have nominated service from the Milk Marketing Board's A.I. service. This meant that, for the correct fee for any given bull, it was possible to have selected cows served with semen from the bull named. He left the bull selection to me so, after going through the 'bull book', I nominated all our best cows, and some of these were very good considering their cross bred ancestry, to be served by a young bull called Snow Regal OIympia. At that time his fee was at the basic rate of £2.50 a straw and I took out an option on 50 straws per annum for two years at the same price. By the time the two years were up his fee had risen to the maximum at that time of £10.00 a straw and there it stayed for the rest of his career.

Bramerton Jane, who learnt how to get more grass

I learnt a good deal more about bovine intelligence while we were at Foxley. Bramerton Jane was a very good cow who had come from a pedigree herd, and she learnt how to get more grass than we wanted her to have. We strip grazed the grass in paddocks. Each of the two main grazing meadows were divided into four strips running the full length of the meadow with an electric fence between each. We then subdivided each strip using a cross wire, and giving the cows sufficient for each feed, moving the fence twice a day. This worked very well,as they cleared up all that they were given and little was wasted by trampling. Jane considered that we did not give them enough, so she worked out a system that would allow them to have a bit more, and it took all my ingenuity to stop her. We started finding bent stakes along the cross wire with grazed grass beyond them. We thought it accidental at first, but soon realised that it was not when I went to get the cows up one afternoon and found every stake bent over at the same crazy angle. I moved the wire, straightening each stake as I did so, and when I turned out again that evening I kept watch. A very short watch revealed the culprit, as Jane went up to the first stake and put her head against it below the insulator and bent it over, had a mouthful or two beyond it and moved on to the next, to repeat the performance. By this means, the cows as a whole were able to 'steal' another two feet of grass right across the pen at each feed. I finally beat her when I put a second insulator low down on each stake and ran a wire between the two insulators so that the whole of the front of the stake was effectively 'live'.

Then there was the case of Elstoke Eclipse Sunflower, as stingy an old cow as I have ever come across. I still have a semi numb spot on my right thigh where she lashed out and caught me a piledriver nearly 30 years ago! I learnt to live with her, though it was always advisable to have her leg strapped for milking. Ted hated her when he was relief milking, and the feeling was mutual. She normally came in about halfway through milking to any stall which happened to be empty. One morning, she came in last and made a beeline for the stall nearest the dairy door. Milking finished at a few minutes to seven thirty, and at about that time, Ted came in to get the milk for the house. An elderly lady came in two or three times a week to help Mrs Littlewood in the house, and she had a bottle of milk from the farm each day. Ted always brought the bottle in, said 'Morning Alec!' and rinsed it under the tap over the washing down tank just inside the parlour. That morning, Ted came in, said 'Morning'-Wham! Old Sunflower's foot went back like a sledgehammer, shaving Ted's knee by a whisker. Ted swore loud and long. The next morning, we had a repeat performance. On the third morning, Ted rounded on me and accused me of deliberately getting Sunflower in near the dairy door last. Naturally, I denied the charge, but thereafter I had to go outside at some time during milking and drive her in so that she was gone before Ted put in an appearance, in order to keep the peace between us. It is no good telling me that she worked that one out by instinct because I know full well that she did not!

I have not mentioned one of the nightmares of a cowman's life, especially on a mixed farm. That is, of the cows getting out at night into a field of sugar beet. Unless they are properly wilted, even sugar beet tops are poisonous to cattle, and if still in the ground, they are deadly, if enough are eaten. On one occasion, the Cranworth herd got out early in the night and stayed on a sugar beet field until we went to get them up at five o'clock next morning. Every one of the cows had to be drenched, though I cannot now remember what it was the vet gave them, as treatment has moved on since that time, the last treatment that I gave them for that condition being a 400 ml bottle of PMD20% milk fever injection containing magnesium and dextrose. On the above occasion, in spite of the drenching, a dozen cows were seriously ill and two of those died. We had such an occurrence at Foxley, in which several needed treatment, but it was the getting them back to the farm that was the biggest problem on that occasion. It was a brilliant night, or early morning in October. There was a full moon and a sharp drop in temperature had caused a head high fog to form. It was absolutely clear over head, but on the ground, one could not see ones hand in front of ones face-more or less. The cows got out onto sugar beet belonging to a neighbouring farmer, had a feed, and then wandered from there into a ten acre field of big old black currant bushes. I will leave it to the reader's imagination as to the job we had, after I had summoned help, finding sixty odd poisoned cows in ten acres of black currants in bright moonlight with dense fog. It was daylight before we got them home.

On one other occasion when the cows got out in the dark, the result was rather amusing. There was a gangway across the end of the meadow nearest home, which the cows had to follow to get to the further strips. The strip furthest from the gate on the first meadow was bounded by the wall of the Old Rectory. For part of the time we were at Foxley, the person living there was Brigadier General Freeland who had moved there directly after a period as C.in C. Northern Ireland. As a result of this, he had a twenty four hour police guard. The morning after my day off one week, Ted came in as usual.

"Morning Alec." He paused. "Did you hear anybody singing up your way yesterday morning early?"

"Singing? No! Should I have done?"

Ted tipped his hat back and scratched his head. "Well, that's a funny thing, but I never heard anything either, and I was over on the pen near the rectory and all, and I never heard a sound."

A sudden suspicion crossed my mind. "What were you doing on the pen by the rectory? The cows were on the lower meadow the night before last."

Foxley herd strip grazing, the Old Rectory in the background

"Yes I know, but I went along there last thing with the tractor and must have forgotten to put the gangway wire up when I came back. Anyway, them old buggers they twigged it, and all went haring along to the rectory paddock with me after them. Did I call them some names? Then about ten minutes after I got home and started milking, this policeman came into the parlour and asked me if I'd heard anybody singing 'Danny Boy' across the fields. I told him, I never heard anything." Just then, George came in. He lived in the council houses opposite the rectory. Ted turned to him. "George, did you hear anyone a singing on the fields yesterday morning just after five?" George had not. By now, I was having the greatest difficulty keeping a straight face, having heard Ted in full cry in the past. I said nothing then, but two or three days later I remarked tongue in cheek, as it were.

"Somebody was telling me that Lady Freeland was very upset when she heard all that shouting and swearing just over the wall. She thought that the Irish had arrived."

Ted's face fell a mile. "She didn't hear me did she?"

I hid a grin. "I think that's why the policeman came over."

He went off looking very crestfallen.

He was very extrovert when it came to protecting his crops. I suppose the era of the universal banger had arrived by the early seventies, but it certainly would not have been as efficient as it is today. In any case, Ted preferred the old and well tried method of having a sheet of corrugated iron staked up just inside a newly drilled field, and every so often, going into the field and belabouring the captive sheet with a big stick and at the same time screaming "AAAAAA-AAAAAHHHHHHHHHHHHH!" for as long as he could maintain breath. It kept the Rooks and Pigeons on the move if it was done often enough, and it let the residents of the third parish down wind know where the farmer was. During the course of one winter, we had a plague of Sparrows on the farm. Ted always grew a large acreage of wheat, and to store it, he had an asbestos barn containing six or eight siloes that were open at the top. The drawback was that the barn doors had to be open for much of the day as tractors and workshop tools were kept in the other end. The Sparrows must have come from several parishes around to fly up onto the top of Ted's silos and enjoy a jolly good feed of plump wheat kernels. Although in the middle of the village, Ted finally resorted to the sheet of corrugated iron and stick to frighten them off. I rounded the corner of the buildings one morning just as "AAAAAHHHHHHH BANG! BANG! BANG!" came from the open barn doors. I stopped in astonishment as Sparrows came pouring out through the doors, through the ventilators all along the centre of the roof, and even one or two through a hole in a pane of glass. All looped up and landed on the asbestos roof. There were so many that the whole roof was grey with them. I estimated there to be at least a thousand. Ted ran out of breath and the banging subsided. After about half a minute, one or two Sparrows started a tentative chirping, as only Sparrows know how to, and all the other 998 followed suit. "AAAAAAAAAHHHHHHHHHHH BANG! BANG! BANG! BANG!" Thoroughly demoralised now, the Sparrows leapt skywards as one and disappeared over the houses in a great grey cloud, apparently heading for Foxley Wood, half a mile away. Silence, this time unbroken. After about a minute, there came a muffled voice from the depths of the barn.

"I'll give you bloody chirrup!"

One of my duties at Foxley was to see to the ordering of such dairy supplies as I had need of. This was done through the various reps, of which, at that time, there seemed far too many. Some days, they would seem to be calling almost continuously, and we only actually did business with two or three. Ted hated Reps. He told me soon after we went there,

"I only want to see Crocky. Any others that call that we don't deal with, get rid of them. I don't care what you tell them as long as they go."

Crocky was Mr Croxton, representative for R.J.Seaman the local cattle feed merchants, from whom he bought protein supplements to go in the cow and pig feed, and to whom he sold all his corn. On occasion, Ted would come hurrying round the corner and gasp "Rep coming" and would shoot in behind the grain dryer until such time as I was able to move him on. At that time, the mineral reps were the most troublesome. It seemed as if they had only just got round to telling farmers how important it was to feed minerals to all classes of farm stock. Whether this was so or not, Ted had never bothered to feed free access minerals until the year after I started working for him, and only then because, quite suddenly, we had a problem. Cows became unthrifty and started biting the woodwork round the yards and things like that. The vet took a dozen or so blood samples and the result came back showing a deficiency of phosphorus in the blood in all cases, as well as being low in some trace elements. Never mind, there were plenty of reps queuing up to sell us their products. There was the objectionable chap from 'Oasthouse'(using made up names) who I could not stick, so I never patronised him. There was the young chap from 'Minilicks' from whom I'd had minerals whilst at Cranworth, so I had some from him for a few months, and then he came round selling 'Supamin'. "But!" I said, "Last month you told me that Minilicks were the best on the market. Now you say that 'Supamin' are."

He looked me straight in the eye.

"Oh yes! Minilicks are the best on the market, but Supamin are just as good, and they pay me better."

A few months later he was back with the news that three or four reps from Supamin were forming their own company to market a new, urea based feed block, but would still sell Supamin minerals. They did not, of course, but soon started marketing their own 'specially formulated' brand. However, I had formed the opinion that 'Supamin' minerals were the best for our herd, so I switched to the 'new' regular rep, who was actually marketing his own brand of a new liquid feed for cattle, a second subsidiary of 'Supamin'. This time, the association with 'Supamin' was maintained and proved of great benefit to me at a later date and in another herd which also developed a mineral problem as 'Supamin' (whatever their name is or was) were the only firm left in the market at that time, who would actually produce a mineral mix to a customers specification to meet a particular deficiency need.

Alas, having settled in nicely at Foxley and having started to improve the herd very considerably, and looking forward to a long stay, things turned out very much otherwise, thanks to the march of

Foxley herd, July 1967, just having crossed the main road

'progress'. The powers that be had been pressed for years by the residents of the next village, Bawdeswell, that they wanted a by-pass. Foxley and Bawdeswell are both on the A1067, and I must admit that it was a regular nightmare, even in the early 70's, especially on a foggy morning in October, getting the cows across the main road for morning milking, on my own, and with no warning lights, and still perhaps an hour and a half to daylight. Many people who knew the road and travelled it at that hour, would be on the look out as they rounded the bend a hundred yards from the crossing, but there were obviously some travellers who were not regulars. When it was clear, it was possible to see the lights of cars leaving Bawdeswell in one direction and Bintry in the other-open the gate and rush the cows across before the cars got there. If it was foggy, I had to push them across as quickly as possible, saying a little prayer as I did so. We never had any problems, though whether by luck or Divine Intervention I do not know, but we very easily could have done. Anyhow, the Bawdeswell by pass was approved. One of those ridiculous routes that bypasses the main village but cuts the lesser neatly in two, so that current signposts are to 'Foxley West' on one side and 'Foxley Street' on the other.

Moreover the present road passes close to the farm buildings, with all but an acre on the far side of the main road. Various possibilities were looked at. There were three options. A flyover for the cows. 'Not possible as the buildings were too close to enable the flyover to be high enough for high loads to go under.' An underpass for the cows. 'Not

Bawdeswell by-pass being built, cutting Foxley in two. September 1973

possible due to the high water table, which would mean that it would be half full of water for much of the year.' Take the cows across the road as we had been doing. 'Not possible due to the speed of the traffic on the new road, with a bend nearby in one direction, making too short a braking distance.' End of story. Sale of all cows and equipment and yours truly was made redundant. Still, I had several month's warning, advertised and received a choice of replies, so I will deal with that part in due course.

Just before we left Cranworth, things had been set in motion which were fundamental to the life and work of the cowman. When we moved to Foxley, things were still being done as they'always'had been, but not for long as everything was being modernised and speeded up on a national scale, and the chaps on the ground had to modernise or get out. First of all, the cleaning and sterilising of milking plant. This had not changed much since milking machines had taken over from hand milking, but having to take it to pieces weekly and scrub it was time consuming, and keeping a coke fired boiler going to provide hot water and steam for sterilising was expensive. The manufacturers of milking equipment and dairy chemicals came up with various solutions which were put into practice and have remained basically the same for the last twenty five years or so, give or take a few modifications. Labour saving refinements are frequently coming onto the market, making the cowman more of a computer operator than a stockman at the present time. One of the greatest changes in the average set up was probably the first. The steam chest was done away with and a 'wash line' was put in parallel with the milk line, and after milking and a first rinse through with cold water to remove surplus milk, a morning hot sterilising solution was circulated, originally for twenty minutes, but latterly frequently reduced to ten, followed by a final rinse with clean water to get rid of the chemicals . The evening sterilising solution was cold water containing hypochlorite. However, this resulted in a build up of 'milk stone' especially in the glass pipes now used almost universally to replace the old stainless steel, and also in the glass jars into which the milk first goes on its journey from the cow to the dairy. This was overcome by a monthly cold circulation of an acid known simply as 'milk stone remover' followed by the regular morning hot water circulation containing a solution of caustic soda. The ultimate step soon followed and was initiated by the Milk Marketing Board, modernisation of the milk collection service which started in about 1965. No more lifting of heavy churns anywhere in the country. A bulk tank was put in the dairy at Cranworth a while before I left, but it was another couple of years before the area which included Foxley was converted to bulk collection. Since that time, and especially since the introduction of milk quotas, the number of herds in the eastern counties has dropped dramatically, and the milk collection industry has contracted sharply and become much more centralised. The village creameries disappeared years ago. Intermediate sized sites such as those at Attleborough eventually succumbed and our milk went either to Harford Bridges on the outskirts of Norwich, or to Kings Lynn. Even this has been superceded. The several contractors who hauled the milk to these creameries were taken over by the Milk Marketing Board about 1990 who brought in even bigger tankers to replace those then in use, and Harford Bridges was closed except as an off loading depot where milk could be transferred from small tankers to bigger ones which hauled the milk to wherever it was wanted for processing. Since that time, the Milk Marketing Board itself has been abolished and replaced by 'market forces'. In practice, Dairy Crest, owners of the Fenstanton processing factory near Huntingdon now have a total monopoly in East Anglia having bought out Long's Dairies of Gorleston , the last independent milk retailer in Norfolk in late 1998. But I have jumped ahead a bit.

The sale day arrived for our 60-70 cross bred Friesian cows and followers plus equipment which included a relatively new bulk milk tank. By 1973, more than a dozen Snow Regal Olympia daughters were

Hag Cherry, one of the young cows, just before the sale

in the herd and milking well, with a further goodly number among the heifers that were coming along. Most of those dozen topped the £200 mark which was top of the market range for commercial dairy cows at that time. Several of the buyers asked my advice and one buyer actually took a dozen of the cows from the herd. We had recorded the milk on a weekly basis for our own benefit, though not under National Milk Records, but nevertheless carefully kept, so buyers were able to make an accurate assessment of each cow both as to looks and performance. Nevertheless, I heard later from two sources that the cows bought had not lived up to their past records and the new owners were wondering why. The answer may lay in a remark that Ted made to me on one occasion.

"I can't make it out. When your predeccessor was here, I always used to get more milk than he did on his days off, but since you have been here, the cows are always down on your day off."

I suppose I shrugged at the time and made no comment, but after that time I took note with a variety of relief milkers, and found that, on average I got between two and ten per cent more than the other chap. They did not skip milking, or anything like that. My own feeling has always been that, with smooth and efficent milking and an affinity with ones charges, it is possible to influence production simply by who is milking. For instance, jumping the gun a bit, as I got nearer retirement and had a bit of land of my own, I only milked for four days a week and the relief man milked for three. Milk recording was done on a daily basis, so a direct comparison was possible. There was very little difference in the time either of us took to milk the 90 or so cows, and by studying the records, it could be seen that a nucleus of the cows gave approximately the same for the whole week. Equally there were a number, usually well over half in any one week were giving up to 10% less when the relief was milking. For example, a cow which was giving 20 to 22kg (yes, we went metric. No more gallons at the source of the milk though pints are still delivered to the doorstep) while I was milking, only gave 18 to 20 kg when the relief was on duty. No good asking me why because I do not know. Put it down to 'cow psychology'.

The cows are sold and we are on the move again. We realise now that it was as well that it happened when it did as, within two or three years, Ted had sold the farm and taken early retirement. There is now no herd of cows at Church Farm Foxley, or on any of the other farms visited so far during our progress through Suffolk and Norfolk, except the last one that we will visit soon. Before we do that, we must take a stroll through Foxley Wood, one of the gems of the Norfolk countryside.

It was probably towards the end of 1963-before we had a car anyway-that I cycled up to the north Norfolk coast and came back to Cranworth across country through by ways not previously travelled. I came down from Guestwick to Themelthorpe and thence to Foxley en route for Dereham and home. Shortly after leaving Themelthorpe, I passed a scene of utter devastation. What had obviously been a large wood on sticky clay had had, so far as I could see, every large tree felled and dragged out onto a wide roadside verge and stacked. No doubt many had already been taken away by that time, but there were still a very large number of trunks partly screening a vast morass of water filled ruts running back into the hinterland through a tangle of hacked off branches. That was all that appeared to be left of Foxley Wood.

When we moved to the village in late 1967, it was not long before I had made a surreptitious visit into that hinterland to see what plants, if any, were still to be found. What a complex population there proved to be. Quite apart from an interesting ground flora, among the first things discovered was a tree of the very rare Wild Service, several more of which were found later. I wrote to the Norwich office of the then

Sorbus torminalis. Wild Service tree

Nature Conservancy Council and asked if Foxley Wood was an SSSI, as I felt that from what I had seen that it ought to have been. I had had some experience of what criteria were necessary to recommend SSSI status for a site as I had done a survey of Potter's Carr at Cranworth whilst there which had gained the interest of the then Assistant Regional Officer of NCC, Mike Schofield who had come out and gone through the Carr with me, and within a year or two the site was scheduled. Though the criteria have been altered somewhat since that time, the NCC, now English Nature , were able to add an interesting marshy meadow at the end of the Carr to bring the site area up above the minimum necessary for it to maintain its status under newer regulations . To return to Foxley Wood. I received a reply from the new Assistant Regional Officer, Dick Hornby to the effect that Foxley had been an SSSI because of the large numbers of old trees that it had contained, but as these had all been felled, they were considering descheduling it as they had no information on the Flora and Fauna. Dick gave me the address of the owner, Mr Dixon-Spain of Thorney, near Peterborough, to whom I could apply for permission to enter, but I also had to get permission from the Wensum Forester of the Forestry Commission who had now taken the lease in hand. It appears that the former lessee had taken the lease from Dixon-Spains simply with a view to cashing in on all the timber, and had then passed on the lease to the

Forestry Commission rather than having the responsibilty for replanting. By that time, the latter had already planted one large block with Corsican Pine which was totally out of keeping on boulder clay in an area of broad leaved woodland. Those two hurdles cleared, I now had to make myself known to Mr Dixon-Spain's keeper who was also responsible as woodman for the two compartments still in Dixon-Spain control, where the keeper reared pheasants, and there was also an area towards the middle of the wood known as the 'sacred ground'. I understand that many years before, perhaps even before the first world war, a Dixon-Spain ancestor had a retreat there where he spent some time during the summer, and, following his death, his ashes were scattered there.

Jimmy Bartrum I found to be a great character in the true Norfolk country tradition. He had not always been a keeper-cum-woodman, as, in his earlier days he had actually been a cowman, so we had some common interest, apart from the wildlife of the wood. He told me many things about the glories of the wood before the trees were felled.

"Did you ever see any of the Purple Emperors that used to be here Jimmy?"

"Why yis! There used to be a king oak right in the middle o' the wood, an' I used to see 'em a flyin' round the top o' that, as big as bats they were."

Somewhere in the middle of the wood there is a colony of Lily of the Valley, but at that time, apart from the conifer block the scrub was very dense and, search though I might through the 300 odd acres of the wood, I could only ever smell them on warm damp, summer evenings as they were remote from the rides.

"I could smell the lily of the valley Jimmy, but could not find them. How big an area do they cover"

"No, you can't find' em very easy 'cause they're well back from the ride, but I reckon they cover nigh on half an acre."

"Do you think they're native in Foxley?"

"No! Old Clarke, my predecessor, he got some roots from Hockering Wood and brought here. He must've known what he was doin' though 'cause they spread so. I've tried several times to move some to different places in the wood, but not one of 'em ha'-took."

This can be explained by the wide diversity of soil types in the wood ranging from chalky boulder clay to an acid medium loam and including wet areas with springs and a stream-cum-ditch with Alder coppice alongside it, a rare and unusual feature in old woodland. There is also a relatively small area of Small-leaved Lime, another native tree with a long history, though the classic example of Small-leaved Lime woodland is to be found at Hockering, where Clarke got his Lily of the Valley roots from. With such a wide range of soils, the whole spectrum of woodland plants occured from Greater Butterfly and Early Purple Orchids and Herb Paris, right through to the acid areas with massed Bluebells and grassy glades with Pignut and Wood Sorrel.

Bluebells in Foxley Wood

During subsequent years I spent many hours in the wood, particularly in company with the late Eric Swann, joint author of the 1968 Flora of Norfolk, and especially during the winter, when we did a survey of the mosses and liverworts to be found in the wood. On one occasion we came across a tree which puzzled us. Being winter time, we had little to go on, and could not decide whether it was a wild apple or a wild pear, though it did not quite seem to fit either. I asked Jimmy next time I saw him, and had managed to describe to him which of the trees in the 300 acres that I meant.

"Oh that one! That's an apple. In fact, that's a William Gladstone. Old Clarke planted that and several more fruit trees that you'll come across in the wood." Local knowledge so easily lost.

My survey of the higher plants listed over 270 species, a record for a Norfolk wood at that time. Nothing nationally rare, though there were a number of local rarities. Slender Spiked Wood Sedge in its only site in the county. Greater Butterfly Orchid, Orpine, Small leaved Lime, which Swann did not accept as a native Norfolk tree until some years later. There were no fewer than six species of St.John's Wort, the drifts of Bluebells were set off to perfection by opportunist blocks of Red Campion in season. Delicate Windflowers-Wood Anemones, nodding in the April breeze and the full range of Avens, the wild Geums, from the small flowered yellow Wood Avens through the nodding pinkish heads of Water Avens and every size shape and colour between yellow through pink and orange, some with smallish spiky flowers, others with large doubles and occasionally growing out a second double flower from the middle of a previous one, all presumed to be the result of hybridisation between the two.

Why did Swann change his mind as to the status of Small leaved Lime in Norfolk?

In 1974, I took him to see a tree of the species at Honingham, growing on the edge of the former 'flood plain' of the River Tud. Prior to this, he had assumed that the species had been planted in woods in fairly recent times for coppicing. However, the Honingham tree has obviously been pollarded many times, the method of coppicing used in Mediaeval wood pasture to protect trees from grazing stock. This old tree displays the character of the ancient forest trees of myth and legend, with its gnarled and twisted trunk and imagined face peering from among the young branches,though these have not been cut for many years now and, at last, there are a few dead branches appearing amongst the living, so it may not live too far beyond the millenium.

The landowner, who manages his woodland in accordance with the best modern principles of Forestry, and knows a great deal about dating old trees, considers this Lime to be contemporary with a line of ancient Oaks which stretch upstream along the edge of the former floodplain for something like half a mile. All the Oaks would appear to be nearing the end of their days. Many are hollow. and most have lost limbs. All are estimated to have started their growth about 1350, which makes the Lime 650 years old approximately. Since 1974, it has been shown that Small-leaved Lime features in ancient hedgerows and on village boundary banks dating back to Domesday and beyond. Dr Oliver Rackham in his book, 'The History of the Countryside' refers to a 'Lime Province' and shows from pollen samples that this stretches as far north as Yorkshire, so it is safe to assume that it is truly native in Foxley Wood where the only introduced trees apart from Forestry Commission conifers would seem to be some Rhodadendron and a few isolated spruce round the Sacred Ground. Foxley Wood can be traced to the

Domesday survey, though not mentioned there by name. The Manor of Foxle (Anglo Saxon, the woody place with many foxes) was held by one Godric who had been granted it by Edward the Confessor. His chief right was of pannage for 300 hogs. If we take the mean as being 1 hog per acre of woodland we have the fascinating suggestion of the present size of the wood. On the death of Godric, the Manor passed to the Crown until it was granted by Henry II to Sir William de Monte Casino-and so on down the ages. At the time of Henry II the reference was to 'Foxley Park', a term used to denote a royal hunting ground and, it was still known as Foxley Park on an early Ordnance Survey map of the parishes of Bawdeswell, Foxley and Sparham published in 1818, at which time the woodland boundary was exactly as it is today.

Actually, the felling of the ancient trees had started as long ago as 1915 the resulting timber being cut into lengths and sent over to France for shoring up the trenches. There can be little doubt that the process started then and brought to completion in the early sixties was the real reason behind the loss from Foxley and ultimately from the county as a whole, of a complete range of woodland butterflies of which, the Purple Emperor was the most spectacular. Consulting the 'Butterflies of Norfolk' (M.R.Hall 1991) we find mention of Foxley Wood under Silver-washed, High Brown, Pearl Bordered and Dark Green Fritillaries. Of these, the last named is now restricted to a few small heathy areas near the coast, whilst the others are extinct in the county. Whilst leading a party of members of Lowestoft Field Club through the wood one Sunday in 1972, we were fortunate enough to catch a brief glimpse of three Pearl Bordered Fritillaries, and, so far as I am aware, these were the last seen. There may be some truth in the suggestion that the Fritillaries died out due to more intensive pheasant rearing following the retirement of Jimmy Bartrum as keeper, when Alan Lambert became gamekeeper/woodman in his stead in the Dixon-Spain compartments. I would very much doubt it though, bearing in mind that during the 1890s, the accounts of the Evans-Lombe estate at Bylaugh, under whose jurisdiction Foxley was at that time, show that no fewer than ten woodmen and ten keepers were kept in the wood. Ten keepers would not have been twiddling their thumbs. In addition to apprehending would be poachers, one imagines that they must have reared far more pheasants than Alan Lambert ever did. Nevertheless, I will admit that the seed capsules of members of the Violaceae, whether Dog Violets or Wild Pansy are very acceptable to pheasants as food, as I know to my cost, having grown Heartsease for the wild flower seed trade for a while in the past. Having recorded a depleted list of butterflies, some 270 species of higher plants and over 60 species of mosses and liverworts with the help of Eric Swann, what happened next?

I sent the results to Dick Hornby. NCCs senior expert in forest management, Dr. George Peterkin came down from Monks Wood Experimental Station on a number of occasions to study the wood at first hand. On one occasion, he brought Dr. Oliver Rackham with him, who probably knows more about ancient woodland than anyone else in the country. Far reaching proposals were put forward which included the curtailment of the Forestry Commission's coniferisation of the wood which, by that date, had already achieved 25% coverage. A site meeting was held which I was invited to attend, as were officers of the Nature Conservancy Council, the Norfolk Naturalist's Trust, The Forestry Commission and Mr Dixon-Spain himself. The latter actually came and thanked me for my involvement as, he and his family having had a long association with the wood, they would not have liked to have seen it blotted out with conifers though they would have been powerless to stop this happening as the lease which the Forestry Commission had inherited was for 99 years.

Following the site meeting, a management plan was worked out which was adhered to for the next several years. When Jimmy retired and Alan took over, he restarted active coppicing in the Dixon-Spain compartments and this helped the last of the lovelier butterlies in the wood, the White Admiral which is now one of the features of the woodland glades in mid-summer. During the 1980s , with a new government commitment to privatisation, the Forestry Commission had to redraw their frontiers and return a good deal of woodland to private ownership. They gave up their lease of Foxley Wood and returned it to Dixon-Spain estates who, I suspect, had not really got the resources to deal with it under the management plan, it being a long way from their centre of operations. In 1988, after lengthy negotiations necessitated in part by the need to raise the necessary cash to pay for it, Foxley Wood was sold to the Norfolk Naturalist's Trust, now Norfolk Wildlife Trust, who are now endeavouring to manage the wood in a traditional manner using coppicing and gradually removing the conifers as they mature , to be replaced with traditional hardwoods. A success story for conservation for once, and one in which I like to feel that I have played a small part.

--Winds Slowly O'er The Lea

I dug a perfect flint scraper up from the garden one afternoon. The type Stone Age people used for scraping the fat and other gunge off skins before curing them. It was a very nice piece and, as it was in our garden it joined other bits and pieces we have found since we moved here 24 years ago from the tied cottage at the farm.

Rita and granddaughter Susan, named after her aunt, our Sue, who died of a brain tumour the winter before we moved here, were walking across the neighbouring field one day a dozen years ago when the child's sharp eyes spotted a perfect flint arrowhead not 70 yards from

Flint arrowhead and other finds round Hillcrest

our fence. During hot, dry summers during which crop marks appeared in many places, indicating the existence of old sites, we were able to stand in the garden and look towards a low knoll where the crop marks clearly showed the boundaries of small enclosures and hut circles of the Neolithic people who left the arrowhead and scraper laying about (illustrated on the following page). As if to confirm this, Ronnie Nelson, the local amateur archaeologist, came across whilst field walking some years back and showed me a piece of Neolithic pottery he had just picked up on the same knoll. He also pointed out the site of a Roman bath house on the field close to our fence, and said

Crop marks on Crossley Holes. June 1st 1992

that he was not sure where the Roman trackway ran that was supposed to be hereabouts, nor also why he had found no Roman pottery. In turn, I showed him some stones from a layer that I have dug through on occasions, in the vegetable garden, which appears as if it might have been the floor of a cobbled yard. I failed to tell him that the Roman pottery was to be found under the rose garden as it then was! Our predecessors in this bungalow had found a Roman coin in the garden which they gave to Norwich Castle Museum, to add confirmation of the presence of that group of invaders locally. Another person who has permission to metal detect on that farm and works with the Archaeological Unit from time to time, showed me some of his finds on one occasion, and told me of others. These were mostly coins, and in some cases, half and even quarter coins. In Mediaeval times, currency was made in such a way that a person wanting a half penny could simply break a penny in halves, and a farthing was a halfpenny reduced even further. The coins he had in his wallet that morning included one or two from reigns as far back as Henry II, and a silver penny of Elizabeth I was probably the most valuable, while one of the more interesting was a medallion of the reign of Edward VII. The facial side was typically a farthing, but when turned over, the reverse carried a full inscription of the Lord's Prayer. More recently the same man told me that he had found part of a Saxon horse bridle, the metal part that is, on the field, so there are not that many gaps in the life of the field, part of which had been bought in 1936 or 37 by the then Assistant District Surveyor for the old Mitford and Launditch RDC to build this bungalow on. That that gentleman knew what he was doing, we have no doubt whatsoever. A lovely site on the side of a hill, sheltered from the north, with a view over the valley of the River Tud, and the nearest neighbour nearly half a mile away. He had a well dug which reached down to the water table on the chalk, just thirty feet deep here and, though electricity did not reach this outpost of civilisation until 1963, I would doubt if the early residents minded that much as there were so many other compensations.

The only major gap in the occupancy of the field would appear to have been between Neolithic and Roman times, apart from which,man would seem to have been around more or less continuously. Strangely, apart from the bungalow, there is no other trace of settlement visible on the surface, and very unfortunately, 1998 marked a new tenancy of the farm to which the field belongs, the first crop being potatoes after almost ten years as setaside. The potato crop necessitated the deep ploughing of the field which cut right through the floor of the Roman bath house which was little more than a foot beneath the surface, and one of those machines was used which sorts through everything in the soil bigger than a walnut and deposits it in a line to one side of where the potato harvester will travel. The result of this can only be that any remains that there might have been which could be of archaeological significance have all been churned up together. For instance, it will now be very difficult for a field walker to sort out worked flints from those which have been smashed into bits by the stone machine. Such is progress.

The blood curdling scream of a vixen moving along the valley has just come to my ears above the tapping of the typewriter, to remind me yet again that there are many things that we neither know, nor begin to understand about the world around us, and especially of the animal kingdom over which we arrogantly claim total mastery to the extent that we think we can alter crops - or domestic animals - by gene manipulation and think that there will be no repercussions. Theirs is perhaps a civilisation beyond our comprehension, which we may occasionally catch a glimpse of, if we learn to use our eyes and ears.

When we left Ted Littlewood, the week following the sale, we moved to a tied cottage at Berry Hall, Honingham, where I was to take charge of the Honingham Herd of Pedigree Guernseys, without the benefit of previous Channel Island experience, I might hasten to add! The herd had been decimated a year or two previously by an outbreak of Brucellosus, formerly known as Contagious Abortion. An eradication scheme was now in place, and all infected cows had to go for slaughter, and tests carried out on the remainder over a period of time, until a certain number of tests had proved negative. That time had now been reached and the previous cowman had decided that he did not want to get involved in building up the herd again, to a higher level than it had been before. Under his care, and before Brucellosus, the herd had reached just over 50 but the optimum sized herd to be accomodated in the then three year old dairy unit was 78 milkers. Herd numbers had fallen back to 30, so some stock had to be bought in to make a reasonable showing of getting to the optimum number in the shortest possible time. Mr Meynell, the owner, was not a farmer at all, but a Chartered Accountant, and he had taken over the running of the farm from his elderly parents for whom it had been something of a hobby. Captain Meynell had been in the Navy during the first world war, and I think I am correct in saying that that was his career. Mr Meynell had been a Captain during the second world war, though he never used the title. The herd had started originally from a single house cow called Honingham Iris, who was still in the herd, and stayed on for a few years after our arrival, before she departed in ripe old age. In 1970, a purpose built dairy unit based on that then in operation on the 'Farmers Weekly' model farm was erected about 300 yards from the original farm buildings, and strangely, it was after this had been built that the Brucellosus outbreak occurred and plans had been set back by two or three years. Now we were again on the road to expansion. To the thirty were added eight from a herd dispersal sale in Gloucestershire, cows which were bought by arrangment with the auctioneer, and without anyone from this end having seen them. Mr Meynell and I then drove down to Kent and bought all the cows from a Guernsey herd near Wrotham which were due to calve during the next four to six weeks. Both groups were very mixed from the point of view of quality, and some had a very profound effect on the herd as it would be in twenty years time. From Gloucestershire, the Lavenders, Russetts, and to a lesser extent the Joans made their mark, whilst the Kent cows included Barbaras, Topazes and Lavinias and, dare I say it, one or two that were rather poor. Within a couple of months of our arrival, the thirty cows had grown to the mid-fifties and we were away en route to the mythical perfect herd of 78 milkers.

I soon discovered that a non-farming accountant had a very

different way of looking at his herd to that which I had been used to. For instance, when calculating how the herd was to proceed, there was a forward projection of performance and numbers which predicted that 78 would not be achieved for at least five years, given successful breeding of the 56 in the herd at the beginning of 1974. This was because the built in projections showed that 20% of the herd would be lost from one cause or another every year. I was appalled at this figure at the time, but came to realise after a while that the Guernsey is far less hardy than the ubiquituous Friesian. For instance in some years as many as 20% may get milk fever at calving. Of these,you can be sure that one or two will get it so badly that they will suffer liver damage and will probably never get up once they were down, no matter how many bottles of calcium are pumped into them. For some reason, the incidence of milk fever fell somewhat over the last four or five years of my time as herdsman, with a corresponding drop in the number of milk fever related deaths. At the same time, there was a period with a higher incidence of Acetonaemia, a condition related to energy intake in recently calved cows. Such animals appear off their food, with a drop in milk production and a slightly subnormal temperature. If the cowman has no sense of smell he may be left wondering what is wrong. The clue to the cow's condition lies in her breath which usually smells of acetone, or nail varnish remover. Clinical tests of milk or urine with a colouring agent rapidly changes from white to bright pink confirming the condition which responds quite quickly to a small injection of cortizone. If not treated promptly the animal may develop a milk fever like stagger, eventually going round in tight circles, frothing at the mouth and licking herself feverishly. In such cases a 400ml injection of CMPD 20% is an added aid to recovery, though it is the dextrose that is the required ingredient rather than the calcium or magnesium. My only previous experience of this disorder had been back in the 50s at Timworth, before the days of a simple injection. On that occasion, the vet made up a solution of a whole packet of glucose in water and injected it into the vein with spectacular results. Seeing so much of it in the Guernseys, I soon knew when to reach for the cortizone bottle.

However, a catalogue of veterinary ills does not make very interesting reading, or lurid accounts of difficult calvings, and how poor old Fern jammed on her oversized front teat one day while getting up in the meadow,and I found it laying there the next day.

A bit about the 'model dairy unit' which was my new empire. The dairy unit consisted of a milking parlour and dairy with wash line and bulk tank, the parlour being of the 'herringbone' type in which the cows stand in a staggered line with their heads in their feed bins and their rear ends at the side of the milking pit. Thus, instead of having to bend down for every milking operation, the cows udders were roughly at chest height, so there was no bending at all, and we had four cows on either side of the pit. For the first several years, we had one line of units down the middle of the pit which were changed from side to side as each four cows were milked. Concentrates to provide the cows with their 'feed according to production levels' arrived in the form of manufactured dairy nuts delivered in bulk and blown straight from the lorry into storage bins in the loft over the parlour. Each standing was served by an electrically driven auger at the bottom of one of four outlets from the overhead hopper on either side. The milk was recorded on a daily basis and, once a week the average of the daily records was arrived at, this being transferred as pounds per feed onto a feed chart. The feed chart was fixed in the milking pit above the four switches that operated the augurs on both sides of the pit. As an example, Honingham Russett 10th was currently giving 6 gallons per day, of which, it was calculated that two gallons was being provided for from silage and other feeds, and for the other four gallons, she will need four pounds of nuts per gallon or a total of 16 lbs of nuts per day divided into two feeds. Turn the knob on the feeder to 8 and switch the switch in the middle of the dial to right or left, depending on which side of the pit Russett is standing, and 8 lbs come tumbling down in front of the waiting cow. Apart from cleaning any muck off the outside of the milking rubbers, jars and glass pipes, and giving a turn inside each teat cup with a rotary brush whilst preparing the milking units for circulation cleaning, the only elbow grease which had to be expended on keeping the plant clean was done at six monthly intervals when the actual milking rubbers had to be renewed, whilst other rubber pipes, junctions in the milk and wash lines, etc., were only renewed every second year. It sounds as if the cowman now had very little to do, but nothing could be further from the truth. Now he had to do all the feeding, and for much of the time, all the mucking out and littering as well. This has become the norm in smaller herds, whilst the factory farm type herds may have anything up to 500 cows milked through a twenty standing parlour, with three men responsible for doing everything. A far cry from our early beginnings, with one man to every 15 or 20 cows!

During the summer, the cows were grazed in paddocks much as I had been used to at Foxley, though the paddocks were of one acre and there was no cross wire, during the summer. During the winter, they spent the whole time in the yards and cubicle house, the latter having replaced the principle of the covered yard. The cubicle house had four lines of cubicles totalling the magical 78, one for each cow, which were at a level five or six inches higher than the gangway along which the cows walked. After a year or two, we discovered that we had a bit of trouble with cows slipping on the concrete floors whilst trying to get up, and getting themselves into difficulties. We got around this by installing rubber mats for each cubicle. This may sound like an expensive luxury, but the rubber mats were simply lengths of old conveyor belt from collieries, that had become redundant. These were of about an inch thickness of layered rubber and fabric and were exactly the right width to fit the cubicles. They were also bought cut to the correct length. In addition to preventing slipping when getting up, they also protected the cow's knees from bruising as they lowered themselves to lay down. Straw was provided on the mats, though this was only a fraction of the amount needed to litter a covered yard. The gangways between the two double rows of cubicles was cleaned out daily, much in the same way as had been the cowshed at Garboldisham, though now the job was done with a scraper mounted on the back of a tractor and pushed outside. Here, we come to a design fault which was to have a very profound effect on the herd much later on. Although Mr Meynell had been advised by a friend who was an officer of the Milk Marketing Board, when the unit was planned, no effective means of slurry disposal was included in the plans. It was assumed the number of cows would peak at 78 and not rise higher. With intensive management of the grassland, a proportion of the paddocks would be mucked and ploughed each year and then reseeded by rotation. So, for the first several years, the muck from the cubicle house was pushed out into a corner of the open yard and stayed there until the liquid had drained out. Periodically, the second man on the farm would have a half day or a day carting the drained slurry by tractor and trailer to make a proper heap in the stackyard at the old buildings or, if a paddock was available, it might be loaded directly into the muck spreader and spread directly onto the land.

This saga will be continued in due course. This may sound as if cubicles were wonderful. They were very good, yes, but there is a snag in everything. If a cow went down with milk fever in a cubicle, her situation could be very serious indeed, especially if she got it badly. If she lay still and kept in her cubicle until the injection of calcium had had time to take effect, and was then able to get up unaided, fine. She would be O.K. If, however, she started to struggle, she was likely to get herself too far forward and get jammed up against the rail in front, and would have to be manhandled back into place, which was no easy task.

Ultimately, if she went down in the gangway, or had milk fever badly enough so that she could not easily get up, the only answer was to bring a gate in behind the tractor, manhandle the cow onto the gate, and, with somebody sitting on her head to prevent her struggling off, she would have to be towed out onto the meadow where, with a bit of luck once she had got something on which to get some purchase with her feet, she would be able to get up straight away. Then, she would have to go into a box until she was firm enough on her feet to go back onto concrete floors.

Continuing with the cowman's duties, I also reared all the heifer calves up to six or seven weeks before they went down to the old buildings and, for the first year or two, whilst numbers were still low, I had to look after all the followers that were in the old buildings during the winter as well. In actual fact, with all the modern aids, I had to work far harder during the last 19 years and a bit that I had with the lowing herd, than I did for the 15 years or so immediately previous to that.

The farm itself was very much an oddity in arable East Anglia, being a 100% dairy farm with no arable at all. But then, not many dairy farms of 70 acres supported nearly 80 milking cows plus followers. To expand on the second man working on the farm. In addition to doing such landwork as there was such as muck carting and rotovating the paddocks, he also did the relief milking and spent the rest of his time working in the 30 odd acres of woodland which Mr Meynell has always actively managed, and, in recent years, has won a number of awards for his management.

Even with all my other duties, I still had time, in those early days to look round my herd each day, and to learn a bit more about animal behaviour through the eye of the naturalist.

The home bred cows were fairly lethargic in many ways, contented with their lot and used to everything that was going on. The fifteen from Kent had lived in a different environment. An environment, I suspect, in which they were chivvied along from morning to night seven days a week. There had been 140 in the herd they had come from , and there was a larger Friesian herd on the same farm. I formed the opinion while looking round, that the men on the farm also suffered fairly continual harassment. Those old cows were always on the go. Pushing and shoving each other and everybody else in the herd. Thrusting themselves to the front when it was time to go into the parlour, which meant cake, and first out of the gate when they went onto the meadow.

The Gloucester cows came from a herd which had been looked after by the farmer's wife, and were sold up when she died tragically. They had obviously been made much of and knew that their keeper would look after them in times of trouble. The original Poulton Lavender for instance. I was walking round the yard one day, checking the herd over, when I discovered her hobbling towards me favouring one front foot. To my astonishment, she came right up to me and held her foot up for inspection. I took it carefully and felt in the cleft, removing a sharp flint she had picked up, and which had probably fallen out of the silage. This was a fairly frequent problem, and one of the reasons why we had more foot troubles when the cows were in for the winter than when they were out on the pastures. This was a lesson in co operation between man and beast, and how this can best be accomplished if the herdsperson has the time to take care of his stock by going beyond the barest essentials. It could be argued that this was natural instinct, but I would say not. The natural instinct would have disappeared in the trauma of the sale and being carted halfway across England, to be dumped down in a strange environment and amongst total strangers, with different bovine companions and handlers that were not of the fairer sex. Lavender must have reasoned out that as I was in charge, I would look after her in the manner to which she had been accustomed, and she was right, at least up to the end of the first six years.

In about 1977, Mr Meynell's son, Anthony, who had been educated at Eton, and had been training to follow in his father's footsteps as an accountant, suddenly decided that he wanted to go into dairy farming, so instead of embarking on a course at an agricultural college for two or three years, he decided to short circuit the system and went and had a year working on a large dairy farm in Brazil which belonged to an uncle. During this year, he learnt all the practicalities of large scale dairy farming and did much of the work associated with practical cow keeping including milking, and had a good knowledge of keeping records etc. He returned to the farm with many innovative ideas of which some were admirable, though others would probably have borne more fruit had the farm and herd been twice or three times the size. Perhaps one of the best was to double the number of milking units. Instead of a single row down the middle, a new double milk line was installed with a milking unit for each standing. This did not double the speed of milking, but it did result in a time saving of around 15% to milk the full complement of 78 cows. To compensate for this, a new leanto was built on the back of the cubicle house, to join onto the end of the calf house which was an extension at the north east corner of the cubicle house. This was supposedly to take up to 12 dry cows, but proved to be simply an extension of the cows' sleeping quarters, the dry cows now going down to the old buildings to join the in calf heifers. This brought the hypothetical herd size to 92 milkers plus dry cows. It might be suggested that 78 and 12 only add up to 90, and to the man in the street, this would appear to be thecase. To the Accountant, this is not so as, at any given time during the day or night, there will be at least two cows not laying down, therefore 90 places will house 92 cows with comfort. The next move was to plan to get even greater return on the investment in extra cows-not bought in cows, I would hasten to add,-as the annual calf crop was now resulting in 20 or more new heifers to come into the herd each year. It was suggested that we should embark on three times a day milking. This I baulked at. Three times milking may do well enough where there is a team, but where there is only one full time and one relief milker, this was not on. I knew that it would be useless to object directly, so suggested that the same result could be achieved by dividing the cows into three groups of high, medium, and low yielders and those near the end of their lactations. These would then have different levels of feeding though the overall feed costs would remain roughly the same as they had been. By using cross gates in the cubicle house, it was easy enough to divide that into two easily accessible groups for the high and medium yielders, the low yielders being deemed to be no more than a dozen at any one time, being housed in the extension. The high yielders were milked first in the morning and last at night, the low yielders last in the morning and first at night etc. Much to my relief, the system worked extremely well and we had very high production levels for two or three winters until the introduction of milk quotas. The maximum day's yield from 90 cows milking was achieved in the early spring before quotas were introduced in which we sent away just over 1770 litres or the equivalent of almost 20 litres per cow or nearly 4 gallons each. A Friesian man might scoff, but this was an extremely good result from Guernseys. In fact, for production , the Honingham herd was regularly in the top half dozen in the country published in the annual National Milk Records Report over a number of years.

When milk quotas were forced upon us and a drastic cutback in production was called for, the accountant's training came to the fore. How could we get the best of both worlds? How could we produce the maximum amount of milk for the maximum amount of time without incurring too many penalties? When it became available after a year or two at reasonable cost, more quota was bought or leased, but even so, given the high quality of the herd, and the high numbers that we were still running, we were always in danger of going over the top so, instead of the cows following their natural length lactations, it became customary for them to be moved around like pieces on a chess board. When milk prices were at their peak, we would be producing the maximum amount of milk that we could squeeze out, until the end of the month when prices were due to fall, by which time we would be well over quota for the time in the quota year. A note would be on the dairy office desk that say, on January 1st, I would have to dry off a list of perhaps fifteen or twenty cows, some of which may not have been due to calve until May or June. Very astute management, I have no doubt, but it had never been my way of managing any herd to put financial considerations ahead of the well being of the cows. Readers may disagree.

In my early days at Berry Hall, I was allowed to make decisions of my own regarding the herd, not a little due to the fact that Mr Meynell was only on the farm at week ends, and not always then. I even had some say in the breeding programme. As a result of this, I selected one or two of the bulls used, and, rather gainst Mr Meynells wishes at the time, ordered a few straws from a bull called Vinorton Monty, and also had a say in which cows he should be used on. The reservations regarding this bull were connected with his daughter's milk yield records, which were not as high as some others available, but he had an excellent butterfat record and was known to improve udder-shape considerably. One of the cows I used him on was Honingham Minuette 2nd, a horrible looking old cow with a terrible udder from which she gave a fantastic amount of milk. She produced a daughter by Vinorton Monty, Honingham Minuette 17th who, in her fifteenth year , when she finally came to the end of her long and distinguished career, still looked more like a cow of seven or eight. More importantly, one of her daughters by a Canadian Guernsey bull Anthony had selected, Honingham Minuette 35th, was a really outstanding looking animal. As a result of this, and in order to promote the herd among the Guernsey fraternity, Anthony decided that we ought to start showing. For two or three years after he had come into the business, he had obtained the services of a YTS student for 6 months each year, and the first of these had been a Norwich girl named Sonja who, for a short while after completing her course, had had an agricultural post, but when this came to an end, she had gone with her father, and claimed at that time to be the only girl in Britain doing cavity wall insulation. Anthony had heard that she had been helping some friends show Jerseys at the Royal Norfolk Show the previous year, so he approached her and asked her if she would like to come and take some of our cows to the Suffolk and Norfolk shows. She jumped at the chance. As I had never had the opportunity to do any showing, and in any case somebody had still to do all the work with the herd, I was a very interested spectator. Let me say at the outset that there is a lot of hard work goes into the preparation and training of show animals, and the person who is doing the training and taking the animals to the shows has to be of the right temperament. They must be firm with the animals, and at the same time must have their full confidence. Having said that, and having helped Sonja at the Royal Norfolk Show for several years, I am only too aware that a lot of animals are taken to shows who never ought to be there, no matter how smart they look.

On one occasion, we were taking part in the parade of cattle in the Grand Ring. Several hundred beef and dairy cattle are assembled in subsidiary rings round the periphery, and when all is ready they follow in a long line into the grand ring itself, starting with the Friesians, followed by Ayshires, Guernseys, Jerseys and so on through the beef breeds and ending with cross bred commercial beef cattle. It takes nearly a quarter of an hour for all the cattle to reach their positions in the ring, with the prime beef breeds occupying the last line in front of the President's box.

Now comes the big snag. The organisers of the show always take this point in time to carry out the final judging of the inter beef team event. The judge has to examine the merits of each team in detail, perhaps going back and forth a number of times before making his final assessment, and while this is going on, all the other stock in the ring is expected to maintain its position. Most do so with very little trouble, but there is always the odd one or two who suddenly reach a psychological breaking point. Standing next to us was an elderly gentleman who had brought two heifers to the show and had decided to bring one in for the parade. She was a family pet, and he and his wife had made much of her. I believe they only had one or two cows in all. We had stood for perhaps ten minutes when the heifer suddenly went beserk. The reader will know how, in times of great stress a highly strung human being will suddenly snap and become a screaming termagent. Exactly the same thing can happen with cattle. On this occasion, I had to toss my halter to Sonja who was able to lead both cows out, and help the old boy with his rearing bucking heifer who it was no longer possible to control. Others came to help and we eventually got her back in her standing with her companion where she quickly settled down again.

The following year, we had a very smart looking cow, one of the Joannas' (ex Gloucestershire Joan) who was of somewhat temperamental disposition and was not deemed to be 'ready' in time for the Suffolk Show where the championship was won by Honingham Minuette 35th. I might add that that worthy loved going to shows and knew exactly what to do and when. In fact, she could be a bit over confident almost to the point of thrusting her handler out of the way and marching off in an 'I know exactly where I'm going' manner. She even went to the Royal Show at Stoneleigh in Warwickshire in her best year and won the Reserve Championship for Guernseys. Back to Joanna. Sonja always took the top cows into the judging ring and, as Joanna was in a different class to Minuette, that was alright. Both won their classes and progressed to the championship class which includes the best cow from all the preliminary classes. Sonja elected to take Minuette 35th, the reigning champion, and I took Joanna. I laid my hand on her shoulder as we set out for the judging ring and was not reassured to feel her trembling. We got her round the ring by keeping her as close to Minuette as was possible to gain reassurance, and Joanna was awarded the Championship. Our troubles were only just starting.

She played up on the way back to the cattle lines and it took some while to calm her. Anthony went and consulted the show steward. Did she have to go into the Grand Parade? The answer was unequivocally yes, or forfeit the Championship. We managed to get her into and round the grand ring, and things went reasonably well until we were ready to return to the cattle lines, then she too went beserk. We had to anchor her with her halter to the nearest permanent railing until the rest of the Grand Parade had gone, and then put a second halter on her so that we could have one each side to hold her down and a steward going ahead to make sure that the gangways were clear in case she got away and knocked some dear little old lady to kingdom come. That was in 1989. In 1991 the Show authorities, chivvied by the Health and Safety Executive, at last adopted a wiser approach by keeping the cattle lines and approach routes to the various rings closed to passers by while cattle were on the move. Even after this, I received a nasty knock. We were leading our cows back from the judging ring having just finished a class for the best three matching cows from each herd. Suddenly one of Lord Blakenham's cows broke away from her handler, charged between me and the cow I was leading, throwing me down on the concrete floor and jumping right over me. I took a very nasty blow just above my knee which troubled me for several months and left me with a determination not to take any further part in showing.

As a result of being on the 'inside' for a number of years, I have come to the inescapable conclusion that a fairly high proportion of stock of whatever class suffer trauma which could be avoided with a little more forethought by their owners. The attitude that because they are outstanding looking animals, therefore they must go, takes no account of the animal's psychological make up. What should be considered is, is he or she of the correct temperament to be able to cope with the noise and bustle, the 50,000 people each day who come and look at them and ask silly questions(sometimes) of their handlers? If the answer is 'don't know', then look more closely at the animal concerned, or the herd or flock from which they come. How are they regarded? Are they a lower form of life which either evolution or the Good Lord, according to your point of view, has placed on earth for exploitation by mankind? What do the livestock think about it? Most authoritites maintain that they do not, that they only behave by instinct. I wonder how long this attitude has held sway? Even in our sophisticated Western society, I doubt if it is as long as many people imagine. Certainly not more than the 5000 years since the old Neolithic hunters lost their arrowheads on the field next to the bungalow. The modern descendants of the Stone Age, the Australian Aborigines and the Red Indians of North America treat the wild creatures with far greater respect than does modern western man. For instance, though the spirits guiding the Red Indians were inevitably greater, they nevertheless believed that the Bison were also guided by their own spirits.

We only see our domestic animals in terms of mutton, beef, pork, milk and eggs. In so doing, we assume that the thoughts of the creatures concerned are only 'eat' and 'sleep'. Having spent half a century in close proximity to cattle and studied their idiosyncrasies, I have come to the conclusion that they have a far higher standard of intelligence than we credit them with having. Intelligence that goes far beyond the levels of blind instinct. Here are one or two examples gleaned from a lifetime of study.

A heifer calf was born to one of the cows. Not more than two days later, she was whisked round to the calf house and the cow was back in the cubicle house. Five or six weeks later, the calf was taken from the calf house to the old buildings and joined her contemporaries in one of the boxes or small yards. In due course, all went out to grass, and the following winter were yarded in one of the bigger yards at the old buildings. As soon as she was deemed old enough, usually between fifteen and eighteen months, she was put in calf by AI and in due course arrived back at the place of her birth in one of the calving boxes at the dairy unit. Having had her own calf, she rejoined the main herd where, for the first time since she was two days old, and providing mother was still in the herd, she met her mother for the first time since her early separation. Within a month or two, perhaps only within the matter of a few weeks, she may have been found as part of a family group of mother and perhaps one, two or even more daughters who tended to stay together within the framework of the herd for most, if not all the time. I put this theory to an expert on one occasion. He did not dismiss it other than to say that there was no proof that it was not a natural instinct bringing together animals of a similar temperament. Similarly there is no proof that it was! In any case, with today's cattle bred from a worldwide bank of A1 bulls, the chances of animals within a family having similar temperaments is much less than it was when everybody among the younger stock were fathered by the same old bull which may have been on the farm for four or five years. Furthermore, I noted that the sisters still stayed together as a family unit after mother has gone to market or the knacker or wherever.

Apart from 'how do they recognise each other after such a long separation?' one may ask 'How do they communicate?'.

Cows say 'Moo'. That is as far as their conversation registers to the human ear. A cow who has been turned out, away from her recently born calf will walk restlessly up and down the hedge near the field gate all day, saying 'Moo' at frequent intervals. Eventually, she may find a gap through which she is able to force her way, and will be off down the track to the farm saying 'moo' loudly all the way. The others will not even turn and look. Two weeks later, the same cow may stroll down to the gate and find that it is not properly fastened. She will put her shoulder against it and it will swing back. She will look round and say 'moo' to her fellows and within seconds, the whole herd will be speeding toward the open gate. The'moo' sounded the same to me, but not to her mates . Example, Christmas morning 1974 when the cows were in the yard one minute with a bolt that had not gone properly home. Somebody gave the bolt a lick or otherwise moved it, the gate swung back and, within seconds, 70 odd cows were swarming up the lane and onto the A47!

When the first lady of the herd, Honingham Iris, was nearing the end of her life, she had not been very well and was laying half way down the cubicle house in one of the stalls feeling very down. I noticed that, as the others went past, most of them stopped and had a quiet word with the old girl, even some of the cows that had been bought in, in much the same way that respect is paid in human society to an ailing elder statesman. Sentimental poppycock? No! Just plain straightforward natural observation. There is a great deal that we do not know with relation to intelligence levels in animals, and especially among farm stock, those being closest to man and therefore possibly the least studied. As a final question mark on the subject, I am going to leave the reader with a sheep tale to ponder. In the early 1980s I took all of our little piece of land in hand, the lower part having been rented to the farm on which to graze the calves during the summer. Part I ploughed up and cultivated to grow certain species of plant for seed for the wild flower trade. This left me with a largish paddock for which, I purchased three young Jacob ewes to graze. Inevitably these increased, and by 1989 there were seven in lamb plus two young ewes which I had not been able to sell. I had always had the lambs in May when the grass is growing at its best, and lambing time was fast approaching. One of the three originals developed a prolapse about a week before she was due, so I had to call the vet. A young vet came out and inserted a plastic gadget into the vagina which, when firmly tied round the ewe's body, should have held the prolapse in place.

It did not. I called the vet again next day, and one of the partners came out. We got the old girl into a brick shed at one side of the paddock and, while I held her, the vet made an elaborate cats cradle with red plastic baler twine which was woven to and fro round her body and tied at strategic points. This effectively held the prolapse inside without discomfort to the ewe and, as the vet had predicted, pushed to one side with no problem when she started to lamb, which she did about a week later.

The truss was put in place at lunch time on a Saturday. That evening, seven or eight hours later, I was hoeing among my wild flowers, for even when growing wild flowers, it is necessary to keep the weeds down. The ewe with the truss was feeding quietly on the far side of the paddock, and the other eight, also feeding, were scattered far and wide over the whole area. Not one of them gave so much as a 'Baa', but suddenly, as at a prearranged signal, all stopped feeding and hurried purposefully to a spot about fifteen yards from the open doorway of the brick shed. Forming themselves into a tight block, with the oldest two in front and the youngest two at the back, they marched as one up to the doorway of the shed then, two by two, each pair spent perhaps thirty seconds peering into the depths of the shed as if to make sure there was nothing lurking within before turning away and making room for the second pair. This was repeated until each of the four pairs had had their chance to see that there really was nothing there, after which they all resumed feeding as if nothing had happened.

This was one totally unrehearsed and unexpected incident over which I have pondered many times and have not been able to come up with a rational explantion. It puts the whole field of animal intelligence and communication into disarray unless one concedes that theirs may, after all, not be so inferior to our own, at least collectively in the community.

If this is so, what right have we to treat our livestock as wholly disposable numbers, often herded in close confinement in buildings of dubious suitability, and treated as one might a set of figures in a computer, erasing one here and there as the whim takes the programmer?

The 'buildings of-dubious suitability' can cover a multitude of sins. The old buildings at Berry Hall are a case in point. The covered parts of the so called covered yards, which were probably Victorian in construction, were not adequate in relation to the open area which resulted in them becoming extremely sloshy in wet weather unless they were cleaned out every few days, whereas it was customary to only clean out such yards at the end of the winter when they should be nearly a yard deep in solid muck. The boxes and associated buildings were ill lit and poorly ventilated. In fact, so bad was one box used to house three or four dry cows that it was usually known as the 'Black Hole.' Then of course, we had the dairy unit which was a facsimile copy of the Farmer's Weekly Model Farm Dairy Unit. I wonder how they got on with the original? We had no end of problems of animal health in the copy! Perhaps the fault lay in the construction rather than the design? I cannot say. The building was of concrete breeze blocks to the eaves on the main building, and with walls of the same to a height of seven or eight feet at the ends, with sliding doors at the north ends of the gangways and metal gates at the south end. The upper part of the end walls above the breeze blocks were of vertical 'Yorkshire boarding' which were boards about five inches wide with a narrow gap between each, ostensibly to give ventilation but the ventilation was insufficient, and when we had a bad outbreak of virus pneumonia among the cows one winter quite early in my stay, the vet ordered that a section of these about ten feet wide should be taken off both ends in order that the air could blow through.

The calf house, built on the north east corner, and later blocked off further from the sun when the dry cow house was added at the southern end, had three windows high up to the east, but, in humid weather, the walls tended to reek and, pressure hose it and disinfect it with Jeyes fluid between batches of calves as we might, especially after the numbers of cows increased, and consequently the numbers of calves passing through, there were very few which got through without some ailment or other overtaking them, whether it be scour or virus pneumonia.

The latter also took its toll of calves in the ill ventilated old buildings when they got down there. Meanwhile there were other more sinister things looming on the horizon on a national scale.

As efforts were made to get maximum production in the years before milk quotas took their toll, and margins per cow became more important than welfare, various feed supplements were used to boost production from silage rather than by feeding more expensive concentrate. In our case, the silage was put in the feeders with a tractor foreloader, and on top of this was added so many shovels full of brewer's grains which had itself been ensiled during the previous summer, in a pit cut into the bank opposite the end of the feeding area. This in itself was a jolly good feed, and to this would be added bits and pieces. Dried sugar beet pulp nuts were also excellent value for money. This was boosted for a year or two by the addition of so many kilogrammes of fish meal, but as this was very expensive, it was replaced by high protein meal, the origins of which we were not really aware of at the time. This high protein meal had been available for years and was the result of using animal offal and meat which had been sterilised at high temperature and had been regarded as totally safe. Somewhere during the early '80s, somebody introduced a simplified method of preparing the offal for use by using chemicals 'perfected' I believe, in the USA. What was not realised at the time was that there were certain elements in the offal that were not destroyed by the chemical treatment which had been rendered safe by the former heat regime.

We now had a regular fortnightly 'clinic' with the vet during which he carried out pregnancy diagnoses, internal examinations on cows that had not come on heat, trimmed feet and saw cows with chronic ailments that sometimes defied treatment. One day, Oliver, our regular vet , said to me, "I had a cow with a new disease the other day."

"Oh, what was that?"

"It's called Bovine Spongeiform Encephalopathy."

"What are the symptoms?"

"They start to stagger as if they were on the verge of milk fever, and may fall down. There is a sudden change in temperament. They lash out without warning, especially at things like doorposts. It seems that they can no longer judge distances and it seems to them that the doorposts are closing in on them."

"It sounds nasty. What is the cure?"

"There isn't one. The brains starts to break down and goes like a sponge."

"Where does it come from?"

"No one seems to know at present, though it may be related to a similar disease of sheep called scrapie."

It was only a few weeks after this that one of the relatively young cows came in, staggered slightly as she reached the trough in the parlour and lashed out violently as I went to wash her udder. A kick bar had to be used to milk her, and she lashed out at the doorpost as she went out. I called Oliver and he came out and watched her, and took various blood samples for the Veterinary Investigation Centre.

That was our first case of BSE. Did our feed merchants use protein supplements containing sheep offal? It was from sheep offal infected with scrapie that BSE jumped the species barrier because the chemical rendering of the offal did not destroy the scrapie prions in the same way as sterilising had done. Certainly the high protein supplement spread on the brewers grains must be suspect. I do not know how many cases of BSE we had before I retired, and they carried on for a year or two after that date and included daughters of some of the early cases, long after the responsible material had been banned in cattle feeds which suggests that the disease can be passed on from mother to daughter, unless, of course, there was a loophole somewhere in the feed chain that was quietly closed at a later date? Certainly I lost a dozen or more cows in the next five or six years from the disease, the carcases of most if not all of which found their way via the knacker to the closed refuse-tip at Attlebridge, above the Wensum valley and upstream of Anglian Water's pumping station for Norwich water supply, causing much concern in later years as to the possibility of seepage from the tip into the river, though I would have thought that a pretty remote possibility in view of the lay of the land.

After a year or two of feeding just brewers grains on the silage, we started using pressed sugar beet pulp later in the autumn and into the winter as this became available from the factory during processing of the current years beet crop. As this was cheaper, it eventually replaced brewers grains completely. The grains ensiled very well, going down tight when well trodden and coming out as sweet as the day they were ensiled. The pressed pulp which is all that remains of the sugar beet after they have been sliced small and boiled to extract the sugar, is much more bulky and does not compress as well as the brewers grains, meaning that the air can get into it more. When we first started using pulp, this was as an 'extra' and the loads were tipped onto the silage pad when one of the heaps of silage had been cleared and was covered over with black polythene sheeting which had already been used for covering the silage. Because the air could get between the pulp particles fairly readily, particularly near the surface, when the sheet was removed it was discovered to have become fairly rapidly covered with a layer of multi coloured moulds which had to be removed. Inevitably in every herd, there are one or two abortions each year, and, since the brucellosus eradication scheme came into being, all abortions must be reported and the cow and discharged foetus blood tested and sections taken to the VI centre. None of these ever proved positive for brucellosus, but in 1990 we had one, possibly two abortions which were strangely diagnosed as being of fungal origin. This resulted in much head scratching as even Oliver could not suggest a cause. During the following winter we had two or three abortions in quick succession, all of which were proved to be 'fungal'. I had an idea. I went to the pressed pulp clamp and got a polybag full of mouldy pulp and took it to a friend who lived not far away, Reg Evans who is a retired pharmacist and an expert on Fungi, being Fungus recorder for the county. He examined the moulds thoroughly under the microscope and isolated several, but told me that the main mould was caused by a fungus of the genus *Aspergillus*, probably *Aspergillus fumigatus* which could cause severe respiratory problems in human beings. This gave me more cause for thought than I had anticipated as I had been receiving treatment for mild Asthma for the previous two years, something unheard of in my family. But anyhow, yes, he thought it could equally well cause abortions in cattle. I took my findings to Anthony, and we carried on as before. We arrived at March 1992 and one morning when I came to uncover the pressed pulp prior to putting some on the silage. As I rolled the sheet back, the fumes hit me and completely took my breath away. I staggered off the heap and it took several minutes before I recovered sufficiently to summon help as I could no longer carry on, so I had a trip to the surgery instead. Feeling better a few-days later I returned to work for morning milking, but that was it. I was surprised to discover that hitherto unsuspected dust particles released when the feed augurs were operated in the parlour were sufficient to cause a return of my symptoms. A further trip to the doctors followed by one to see Mr Harrison, the asthma specialist at the Norfolk and Norwich Hospital ended my career with the lowing herd just five months before my 65th birthday, though this was not the end of the story. My demise was diagnosed as being due to *Aspergillus fumigatus* the fungus which causes the condition known as farmer's lung. We also now know that it can cause abortions in cattle. There followed a further series of more serious abortions which reached double figures in number and the loss of at least one cow. Samples of pressed pulp were found to contain massive concentrations of *Aspergillus*, and some of this fungus was also found in samples taken from the pulp storage area at Cantley beet factory. Expensive litigation was entered into against British Sugar in respect of the abortions, but was doomed to failure as there were no other herds that could be traced that had suffered fungus related abortions. This could be because most other herds using pressed pulp were Friesian, and these are much hardier than Guernseys, but I think, with hind sight that the method we used for storage was suspect and that the spores were endemic in the dairy unit. Even in the big silage pit when it was used for pressed pulp and was covered with clean plastic sheet, there would still have been spores left from the last silage clamp. All other heaps of pressed pulp were stored on the flat, including that used for feeding the dry cows and heifers at the old buildings from which most of the abortions stemmed. These heaps were all covered using sheeting previously used to cover silage. As *Aspergillus* is more usually found in forage crops it is fair to assume that there would be a fairly stiff impregnation of spores on the grubby underside of the sheet when it has just come off a heap of silage, and that there would be a rapid multiplication of spores if those on the silage sheeting came in contact with other spores introduced even in relatively small quantity with the pressed pulp. There was probably a single load delivered to the old buildings with this problem, causing the interaction with endemic spores and the bad abortion outbreak of 1992. Earlier isolated cases, and indeed, my own asthma, were more likely the result of spores from the silage transferring to the pressed pulp via the sheet whilst the pulp was still warm making rapid multiplication of spores inevitable.

Until it became apparent that I was not going to be able to return to work, the second man on the farm had had to carry on as best he could, with Anthony doing occasional relief milking. As the latter now had an executive post with the Channel Island Marketing body, Quality Milk Producers, and was often away. Bob, the relief milker was made redundant and the whole operation put in the hands of a husband and wife team of contract milkers. This looked like the beginning of the end, and so it proved to be, even within the five years that private thoughts gave the arrangment. Soon after arrival, the wife became pregnant and the man could not cope, and they were on their way within six months. A second husband and wife team with two teenage children took over, the offspring helping in school holidays etc. They did quite well and continued showing at the Norfolk and Suffolk shows, but after three years, fate again took a hand. You may remember that the original plans did not take proper account of slurry disposal. You may also remember my comments about the unhealthy atmosphere in the buildings generally which was exacerbated by the fact that the herd was 14 larger than the original plan catered for, the only allowance for which had been the building of the twelve cow extension on the back. The parlour, the feeding areas and especially the calf house were still the same size. A field had been bought a mile up the road which was used for the production of grass and maize for silage, and all the home grassland became permanent. A slurry ramp was constructed at the side of the feeding yard which enabled the slurry to be scraped direct into a high sided trailer and taken for storage to a long stretch of disused sunken roadway which had been cut off from the A47 when this had been straightened in 1969. From thence it was carted for spreading on the field near Hockering by contractors once each year. During the early spring of 1995, the premises were closed down by the Ministry of Agriculture for two or three weeks, following an outbreak of Salmonella that left two cows dead. The Ministry investigation found that the slurry disposal did not comply with the ever more stringent safeguards governing this aspect of dairying, the cost of which can be enormous. The spring of 1999 has seen the sale of one of the top Norfolk Holstein herds, one which has been known for its extremely high cow health standards, due to the fact that an upgrading of the slurry handling operation would have cost nearly £100,000. On July 14th 1995 the Honingham Herd of Pedigree Guernseys went under the hammer. The last herd that I cared for in my fifty years with the lowing herd and the last farm along my journey from which the cows finally departed within a year or two of my own departure.

Nearly at the end of our journey, what effect has this had on the wildlife of the places we have been to? Much has been made of the

My last herd, soon to be sold

decline of farmland birds over the past thirty years from a variety of causes connected with changes in agricultural practice. Losses of nesting and feeding habitat has been discussed while we were at Cranworth. The impact of road rationalisation schemes even on such a small scale as that at Foxley has to be compared with such disastrous slashes across the countryside as those at Twyford Down and Newbury bypass which rode roughshod over SSSI designation.

Toxic sprays wipe out whole insect populations in a variety of crops, denying insect feeding birds the wherewithal to feed their nestlings, whether they be Blue Tits collecting tiny caterpillars, Whitethroats searching for aphids, or Grey Partridges seeking sawfly larvae for their newly hatched chicks. The early harvesting of crops followed by almost immediate ploughing of stubbles and planting with winter corn, leaves no weed seeds to feed the autumn flocks of such birds as Yellowhammers, Reed Buntings and Tree Sparrows all of which have declined dramatically.

And why has the gentle Song Thrush almost disappeared from country districts during the last decade? Possibly from a combination of some of the foregoing, but almost certainly the indiscriminate use of slug pellets is implicated. Crops of winter corn are liberally sprinkled with these nasties by a broadcaster mounted on the back of a high speed garden tractor type vehicle. This is not really the whole story, as many gardeners are just as indiscriminate with the use of one of the most lethal of poisons available to them. It is not suggested that Song Thrushes eat the pellets, but they cannot tell the difference between a good, healthy slug or snail and one that has been poisoned. One aspect of the countryside is not as well documented. How has wildlife been affected by the steady demise of the dairy industry, especially in an arable area like East Anglia? It may be remembered from the beginning of this account that in the 1950s almost every farm had its small dairy herd, and many kept beef cattle as well. Travelling through East Anglia today, it is a rare occurrence to find a dairy herd with its attendant acres of permanent grassland, and herein lies another cause for concern. All those acres of permanent grass were filled with earthworms and things like leatherjackets, the larvae of the Crane Fly family which would formerly have provided food for nesting Lapwing, and for winter flocks of the same, as well as flocks of Golden Plover and Starlings, to name but a few. Watch any area of permanent grass which has been grazed from May onwards and it will be alive with Starlings probing for these tasty titbits on which to feed their nestfuls of squawking young, and who will continue squawking as they follow their parents to and fro across the meadows after they have fledged, if there is something for them to eat. Rooks and Jackdaws also join in probing for this bounty, but with all the depredation, clouds of adult craneflies still emerge from the grassland in late summer, many of these being snapped up to feed the line of baby Swallows perched on the nearby fence, or by the Spotted Flycatchers whose young are in their nest in the ivy on the wall of the nearby farm buildings. Following the departure of the Guernseys from the farm, the land was let out to a farmer from another parish. Apart from some small riverside meadows, the pastures are all arable now. Our three or four pairs of nesting Starlings have left us. A Starling in the garden is now occasion for a special note in our nature diaries. There were no Spotted Flycatchers locally in 1998 or 1999 and Swallows were very scarce. Very rarely, a Lapwing may be seen flying over, usually very high up. When I used to go to and from the cowshed across the pastures on dark winter mornings and evenings, the sharp 'scaap' of Snipe would be heard a dozen or more times in the 600 yards, and sometimes a tremendous rush of wings would denote a departing Woodcock. Looking in daylight, I would note that all the summer's dried cowpats had holes probed in them by Snipe seeking the juicy grubs of dung flies and other similar creatures. Occasionally, a larger figure of 8 shaped probe would reveal where the Woodcock had also been feeding.

Even the disappearance of the muck heap from the disused length of roadway has had an impact on local bird populations. Pieces of straw from the drier parts of the heap found their way into the nests of Starling and Jackdaws. Fragments of maize dropped by the cows when eating their silage were quickly snapped up by foraging Jackdaws, Magpies, Pheasants and even wandering Moorhens which came up from the RIver Tud.

One summer, a pair actually constructed a nest in a bush just above the accumulated pool of juice that had run out of the slurry and reared at least two young in these less than salubrious surroundings.

At the end of our journey with the lowing herd, and all the changes that have befallen the dairy industry and the East Anglian countryside, what is the view beyond the Millenium, now less than ten months away at the time of writing? Can things change any more? Can they get any worse? When the herd was sold, it came as something of a cultural shock to have to buy our milk from the supermarket, having always been used to milk straight from the cow, especially the rich Guernsey milk of the last 20 odd years. Milk kept more than two days used to go sour, even in winter. Now, the 'pure, fresh, pasteurised milk' that is served up in plastic containers refuses to go off even after two weeks! What do they do to it? When it has been carted all the way to Fenstanton, to the processing plant, it is separated, all the cream being taken out. Then the absolute minimum of fat is put back in to correspond to whichever grade it is intended to represent. There is a move afoot to do the same with the protein. The cream that is put back stubbornly refuses to rise to the top because, not only is the milk pasteurised, it is also homogenised, which means that the fat is kept in suspension throughout the milk, and is presumably the reason why it 'keeps' so long.

The battle rages on the merits or otherwise of genetic engineering. We are told that the genetic engineering of crops will lead to the need for fewer pesticides, thus benefitting wildlife. But what about the beneficial wildlife that will be put at risk? Genetically engineered oilseed rape has proved to be lethal to aphids which can damage untreated rape. Bees which collect pollen from the engineered rape have also been found to die some two weeks earlier than bees who do not visit engineered rape. Genetically modified sugar beet will kill the aphids that damage the crop, and in additon, the ladybirds that are the natural predators of the aphids will die. There are species recovery programmes in operation for a range of plants and creatures in the Red

Data Book. Will the new Millenium see a species recovery programme for ladybirds - or honey bees?

The continuing research into computerised technology for the farm is being carried out on various experimental farms including one at the well known research station at Silsoe in Bedfordshire. Here,automatic 'milking on demand' is being pioneered. Early attempts to perfect an automatic milking machine which would put the clusters on to the cows teats without anyone in attendance were unsuccessful because cows are not all the same size and shape, and the clusters did not always arrive in the right place. The result was rather like having a blind man to do the job who had never been in a milking parlour before. However, the problem has now been overcome with the use of lasers which can position the clusters accurately onto the cows teats without fumbling and thus upsetting the cow. The final stage is to get rid of the cowman altogether, and this is being achieved at Silsoe, for the cows can wander into the parlour when they feel like being milked and the laser operated equipment will do the whole job. To guard against the cows getting hooked on the idea of being milked on demand, and coming back more often than they ought, they are fitted with neck collars bearing an electronic tag similar to that used on criminals at large in the community and under curfew. This will ensure that they are refused admittance to the parlour before their scheduled time.

During June 1999, about 150 members of the Norfolk branch of the Country Landowners Association toured the Earl Ferrers estate at Ditchingham, Norfolk, where this system has already been operating successfully for nearly two years. Here, two robot milkers each deal with up to 60 cows and milk them up to four times per day each. The cost of the system was given as £100,000 for the first robot and all the vital electronic softwear and program information. The second robot is much cheaper.

Certainly the 20th century has seen the dedicated herdsman become an endangered species, though I doubt if this will ever lead to a species recovery programme!

Hector was right. I WAS one of the last of the old boys.